ELEMENTS OF PSYCHO-ANALYSIS

ELEMENTS OF PSYCHO-ANALYSIS

BY

W. R. BION

D.S.O., B.A., M.R.C.S., L.R.C.P.

WILLIAM HEINEMANN · MEDICAL BOOKS · LIMITED
LONDON

First published 1963

THIS VOLUME APPEARS UNDER THE AUSPICES OF THE MELANIE KLEIN TRUST

Printed in Great Britain by the Pitman Press, Bath

ACKNOWLEDGMENTS

IT is a pleasure to acknowledge the help I have received from Dr. Elliot Jaques and other members of the Melanie Klein trust notably Dr. R. Money-Kyrle, who made a number of helpful criticisms when he read my manuscript, and Dr. Segal and Miss Betty Joseph. The book was completed before I heard Dr. J. J. Sandler's paper on *The Ego Ideal and the Ideal Self*, or I should have explored the significance of his approach for the matters discussed here.

A debt of a different kind I owe to my wife for her constant help and encouragement.

W. R. B.

CHAPTER ONE

BECAUSE PSYCHO-ANALYTIC theories are a compound of observed material and abstraction from it, they have been criticized as unscientific. They are at once too theoretical, that is to say too much a representation of an observation, to be acceptable as an observation and too concrete to have the flexibility that allows an abstraction to be matched with a realization. Consequently a theory, which could be seen to be widely applicable if it were stated abstractly enough, is liable to be condemned because its very concreteness makes it difficult to recognize a realization that it might represent. Conversely, if such a realization is available, the application of the theory to it may seem to involve a distortion of the meaning of the theory.[1] The defect therefore is twofold: on the one hand description of empirical data is unsatisfactory as it is manifestly what is described in conversational English as a "theory" about what took place rather than a factual account of it[2] and on the other the theory of what took place cannot satisfy the criteria applied to a theory as that term is employed to describe the systems used in rigorous scientific investigation.[3] The first requirement then is to formulate an abstraction,[4] to represent the realization

[1] An instance of this can be seen in J. O. Wisdom's paper on "An examination of the Psycho-analytical Theories of Melancholia", p. 18, where he clearly states the need for an extension of theory, but sees that it involves making a supposition about what M. Klein's view could have been.

[2] In grid terms, too much G3 instead of D or E3.

[3] Too much C3 instead of G4.

[4] The concept of abstraction will be discussed at length; its use in the early stages is provisional. Such a formulation would be in G.3.

that existing theories purport to describe. I propose to seek a mode of abstraction that ensures that the theoretical statement retains the minimum of particularization. The loss of comprehensibility that this entails can be made up for by the use of models to supplement the theoretical systems. The defect of the existing psycho-analytic theory is not unlike that of the ideogram as compared with a word formed alphabetically; the ideogram represents one word only but relatively few letters are required for the formation of many thousands of words. Similarly the elements I seek are to be such that relatively few are required to express, by changes in combination, nearly all the theories essential to the working psycho-analyst.[1]

Most analysts have had the experience of feeling that the description given of characteristics of one particular clinical entity might very well fit with the description of some quite different clinical entity. Yet that same description is rarely an adequate representation even of those realizations to which it seems fairly obviously to be intended to correspond. The combination in which certain elements are held[2] is essential to the meaning[3] to be conveyed by those elements. A mechanism supposed to be typical of melancholia can only be typical of melancholia because it is held in a particular combination. The task is to abstract[4] such elements by releasing them from the combination in which they are held and from the particularity that adheres to them from the realization which they were originally designed to represent.

For the purpose for which I want them the elements

[1] Compared with the tendency to produce *ad hoc* theories to meet a situation when an existing theory, stated with sufficient generality, would have done. Compare Proclus, quoted by Sir T. L. Heath, on Euclid's Elements (Heath, T. L.: *The Thirteen Books of Euclid's Elements*, Chap. 9, C.U.P. 1956).

[2] A consequence of $Ps \leftrightarrow D$. See Chap. 18.

[3] A consequence of ♀♂. See Chap. 18.

[4] See footnote 4 on p. 1.

of psycho-analysis must have the following characteristics: 1. They must be capable of representing a realization that they were originally used to describe. 2. They must be capable of articulation with other similar elements. 3. When so articulated they should form a scientific deductive system capable of representing a realization suppose one existed: other criteria for a psycho-analytic element may be educed later.

I shall represent the first element by ♀ ♂; as I have discussed it at length in *Learning from Experience*[1] my account here will be brief. It represents an element that may be called, though with a loss of accuracy, the essential feature of Melanie Klein's conception of projective identification. It represents an element such that, if it were less it could no longer be related to projective identification at all; if it were more it would carry too great a penumbra of associations for my purpose. It is a representation of an element that could be called a dynamic relationship between container and contained.

The second element I represent PS ↔ D. It may be considered as representing approximately (*a*) the reaction between what Melanie Klein described as the paranoid-schizoid and depressive positions, and (*b*) the reaction precipitated by what Poincaré[2] described as the discovery of the selected fact.

I have already discussed the signs L, H and K, in *Learning from Experience*. They represent links between psycho-analytic objects. Any objects so linked are to be assumed to be affected by each other. The realizations from which they have been abstracted are usually represented by the terms "love," "hate," and "know."

[1] Bion, W. R.: *Learning from Experience*. Heinemann.
[2] Poincaré, H.: *Scientific Method*. Dover Press.

Using the notation R derived from the word "reason" and the realizations it is thought to represent, and *I* derived from the word "idea" and all realizations it represents including those represented by "thought"; *I* is to represent psycho-analytical objects composed of α-elements, the products of α-function. I have described what I mean by this term elsewhere (*Learning from Experience*). By α-function I mean that function by which sense impressions are transformed into elements capable of storage for use in dream and other thoughts. R is to represent a function that is intended to serve the passions, whatever they may be, by leading to their dominance in the world of reality. By passions I mean all that is comprised in L, H, K. R is as associated with *I* in so far as *I* is used to bridge the gap between an impulse and its fulfilment.[1] R[2] insures that it is bridged to some purpose other than the modification of frustration during the temporal pause.

[1] Freud, S.: *Two Principles of Mental Functioning.*

[2] I have not carried through the discussion of R because I do not yet feel in a position to see its implications. I include it because my clinical experience persuades me of the value of such an element and others may be able to use it incompletely worked out though it is. See Hume: *A Treatise of Human Nature*, Book II, Part III, Section 3. Clarendon Press 1896.

CHAPTER TWO

PSYCHO-ANALYTIC theories suffer from the defect that, in so far as they are clearly stated and comprehensible, their comprehensibility depends on the fact that the elements of which they are composed become invested with fixed value, as constants, through their association with the other elements in the theory. This phenomenon is analogous to the phenomenon of alphabetic script where meaningless letters can be combined to form a meaningful word. The elements in Freud's theory of the Oedipus situation, for example, are combined, by their association to form the narrative of the Oedipus myth, and so achieve a contextual meaning that gives them a constant value. As elements in a description of a realization that has been already discovered this is essential to their usefulness: as components of a theory that is to be used in the illumination of realizations yet to be discovered it is a defect because the constant value impairs the flexibility needed.

The abstractions intended to be elements of psycho-analysis should be capable of combination to represent all psycho-analytical situations and all psycho-analytical theories. For this to be true the chosen elements must be essential in the sense described on p. 7. I propose to devote discussion to this topic before pursuing the problem of abstraction[1] of which the solution is so important if the elements chosen, as the elements of psycho-analysis, are to be capable of use in the

[1] See Chap. 18.

construction of theoretical systems. The first step is to consider what phenomena of those present in analytical practice appertain to the elements of psycho-analysis. We may proceed by following three courses:

1. We may search for the elements as their secondary[1] qualities occur and may be recognized in psycho-analytical experience.

2. We may search for the elements as their representations occur and can be isolated in psycho-analytic theory.

3. We may investigate procedures 1 and 2, and combine them as a source from which to abstract elements.

I shall consider first the observability of the chosen elements, although it might be thought, since the elements enter into the composition of all psycho-analytic theories and are essential, that the first need would be to see if these elements can be detected in the theories.

If a patient says he cannot take something in, or the analyst feels he cannot take something in, he implies a container and something to put in it. The statement that something cannot be taken in must not therefore be dismissed as a mere way of speaking. It implies, furthermore, a sense of at least two objects. It might be stated ♀♂ ≥ 2. In certain circumstances, also observable in analysis, the sense of "two-or-moreness," may become obtrusive. For the present I shall ignore the implication of number although the element I wish to isolate cannot be correctly described unless it is understood that ♀♂ ≥ 2.

It is obvious that the number of occasions on which it is verbally stated that something is "in" something also might be innumerable and correspondingly insignificant. The patient is "in" analysis, or "in" a

[1] Secondary in the sense in which Kant uses the term.

family or "in" the consulting room; or he may say he has a pain "in" his leg.[1] Judgement of the importance or significance of the emotional event during which such verbalizations appear to be apposite to the emotional experience depends on the recognition that container and contained, ♀♂, is one of the elements of psycho-analysis. We may then judge whether the element ♀♂ is central or merely present as a component of a system of elements that impart meaning to each other by their conjunction.

Considering now whether it is necessary to abstract the idea of container and contained as an element of psycho-analysis I am met with a doubt. Container and contained implies a static condition and this implication is one that must be foreign to our elements; there must be more of the character imparted by the words "to contain or to be contained." "Container and contained" has a meaning suggesting the latent influence of another element in a system of elements. As the same objection can be levelled against "to contain or to be contained" I shall assume that both statements are contaminated by the presence of elements of an undeclared system of elements (e.g. the latent effect of the model discussed by me in *Learning from Experience*). I shall therefore close the discussion by assuming there is a central abstraction unknown because unknowable yet revealed in an impure form in statements such as "container or contained" and that it is to the central abstraction alone that the term "psycho-analytical element" can be properly applied or the sign ♀♂ allocated. From this definition it is clear that the supposed psycho-analytical element cannot be observed. In this respect it is not unlike Kant's concept of a thing-in-itself—it is not knowable though

[1] Cf. Ryle Gilbert: *Conception of Mind*, p. 22, not a category mistake but an expression of unconscious awareness of ♀♂ as the set to which all belong.

primary and secondary qualities are. Yet it is different in this respect. The phenomena of container and contained are knowable as secondary qualities. The central abstraction is only a phenomenon in that I as an individual am aware that it is convenient for me to postulate the existence of something that has no existence, as if in fact it were a thing-in-itself. If I postulate the existence of a table as a thing-in-itself, I do so because I believe it exists and that its existence is the explanation of the phenomena that I group together in a set with the name "table."

This explanation is necessary because I am anxious to establish the elements of psycho-analysis on a foundation of experience. The element I hope will be an abstraction of the type given by my example of ♀♂, container and contained. It will be a "central abstraction unknown because unknowable" but adumbrated, in an impure form, by its verbal representation. It will have the same status and quality as the object that we aspire to *represent* by the word "line" or a line drawn upon paper, has to the word "line" or a line drawn upon paper.

CHAPTER THREE

The elements are functions of the personality.[1] Of them all it may be said that each is a function of something else and each has a function. In so far as each *is* a function the term "function" has a meaning similar to that with which it is associated in mathematics. It is a variable in relation to other variables in terms of which it may be expressed and on the value of which its own value depends. In so far as each function *has* a function the term "function" is used as the name for a set of actions, physical or mental, governed by or directed to a purpose. Whenever I use the term "function" I use it to denote something which is and has a function. In so far as it is a function it has factors: in so far as it has a function it has aims.[2]

For the present I propose that the elements of psycho-analysis are all without exception functions using the term in the sense I have just adumbrated. The sign representing an abstraction must therefore represent a function that is unknowable although its primary and secondary qualities (in the Kantian sense) are. As I propose to consider the elements as observable phenomena it must be assumed that I am talking about primary and secondary qualities of elements and not the abstractions or signs by which I represent them. What, of all that can be seen in the course of any analysis, are we to choose as the functions

[1] For further discussion of "function" see Bion, W. R.: *Learning from Experience*.
[2] This point will be clearer when reference can be made to the grid. See Chap. 6 *et seq.*

of personality that are also elements of psycho-analysis? The choice is already limited by the criteria I have proposed (Chapter 1, p. 6). We must now limit it further because the element must be a function in the sense that I have proposed for this term and, furthermore, must be "seen" in the course of analytic work. But how are the qualities of elements to be pronounced "seeable" in view of the notorious fact that some analysts purport to be able to see things the very existence of which is denied by others, a disagreement that is common enough between patient and analysand even though they share the "seen" experience?

As a criterion for what constitutes a sensible experience I propose common sense in the meaning that I have given it elsewhere, namely some "sense" that is common to more than one sense. I shall consider an object to be sensible to psycho-analytic scrutiny if, and only if, it fulfils conditions analogous to the conditions that are fulfilled when a physical object's presence is confirmed by the evidence of two or more senses. It is evident that it can only be analogous, for, in our present state of knowledge even anxiety, at least in others, is a deduction. The problem is to determine just how far we can go in accepting deductions from sensa as having, in the field of psycho-analysis, the same validity as sensa have in physics or philosophy. I have no doubt that my impression that a man is anxious has the same validity as my impression that a stone, say, is hard. But for my impression to be valid I consider it necessary to feel the stone to satisfy myself of its hardness and at least to look at it to satisfy myself that what I touch is a stone. The correlation thus established entitles one to claim the term "common sense" to characterize one's view that the given object is a stone: and that the view that it is a stone is common to one's senses and therefore a common sense view, the term "common sense" being used with more than conversational precision. The problem is to establish some similar usage or convention to define the nature of the sense by which we apprehend a psycho-analytical element and, the counterpart of this, to define the nature of the dimensions of a psycho-analytical element. Implementation of this plan seems, as is so often the case in psycho-analytical investigation, to pre-suppose what we wish to discover. In writing this I have to start somewhere and this produces difficulties because the start of a discussion tends to impose an appearance of reality on the idea that the matter discussed has a start. Psycho-analytic investigation formulates premises that are as distinct from those of ordinary science as are the premises of philosophy or theology. Psycho-analytic elements and the objects derived from them have the following dimensions.[1]

1. Extension in the domain of sense.
2. Extension in the domain of myth.
3. Extension in the domain of passion.

An interpretation cannot be regarded as satisfactory unless it illuminates a psycho-analytic object and that object must at the time of interpretation possess these dimensions. In view of the importance I attach to these dimensions I shall discuss each of them in detail.

Extension in the domain of sense need not detain us long. It means that what is interpreted must amongst other qualities be an object of sense. It must, for example, be visible or audible, certainly to the analyst and presumably to the analysand. If the latter presumption turns out not to be the fact the grounds for the presumption must be such that the failure in correspondence must be regarded as significant in itself. Put in another way when the analyst gives an

[1] Discussion of the grid in Chap. 18 and later will explain more fully what I mean by these dimensions.

interpretation it must be possible for analyst and analysand to see that what he is talking about is something that is either audible, visible, palpable or odoriferous at the time.

It is more difficult to give a satisfactory explanation of what I mean by extension in the domain of myth.[1] I cannot conceive the possibility without it of model making as part of the equipment available to the psycho-analyst. Suppose that a patient is angry. More meaning is given to a statement to that effect if it is added that his anger is like that of a "child that wanted to hit his nanny because he has been told he is naughty." The statement in quotation marks is not an expression of a theory in a genetic exposition. It must not be supposed to express a theory that small boys hit their nannies if they are called naughty. It is a statement akin to the type of statement that philosophers contemptuously dismiss as mythologies when they use the term pejoratively to describe bad theories. I require, as part of analytical scientific procedure and equipment, statements of this kind. They are not statements of observed fact, or formulations of a theory intended to represent a realization: they are statements of a personal myth. Unless the experience of the psycho-analytical object is accompanied by a formulation by the psycho-analyst of a statement that has this type of component it lacks a necessary dimension. I shall refer to this dimension as the myth or "as if" component.

I have deliberately chosen the term "passion" rather than what might appear to be more usual terms for the last dimension, partly because the more usual terms have a meaning that should be left undisturbed. By "passion" or the lack of it I mean the component derived from L, H and K. I mean the term to

[1] The problem relates to the discussion of row C of the grid.

represent an emotion experienced with intensity and warmth though without any suggestion of violence: the sense of violence is not to be conveyed by the term "passion" unless it is associated with the term "greed."

It may seem that by introducing passion I am repeating what I have said already by including the L, H and K, as elements. This is not so; by passion I mean one of the dimensions which L, H or K must possess if it is to be recognized as an element that is present.[1] Furthermore the evidence that passion is present that may be afforded by the senses is not to be taken as the dimension of passion. That is to say, if the angry tone of the patient is judged to be evidence of hate it must not be assumed that passion has been discerned as a dimension of a psycho-analytic object, H in fact. Evidence may be afforded by the senses, in such an episode, that can be correlated with the evidence, sensuous perhaps but not sensible, of passion. Awareness of passion is not dependent on sense. For senses to be active only one mind is necessary: passion is evidence that two minds are linked and that there cannot possibly be fewer than two minds if passion is present. Passion must be clearly distinguished from counter-transference, the latter being evidence of repression. Further consideration of passion is not relevant to the immediate issue of passion as one of the dimensions of a psycho-analytic object and therefore of a psycho-analytic element.

[1] Cf. Chap. 19 on "feelings."

CHAPTER FOUR

In the first chapter I said that the development of psycho-analytical practice was hindered through lack of work on the elements of psycho-analysis and gave examples of what might be objects of a search for such elements. In the second chapter I discussed criteria by which objects proposed as elements might be judged, stressing observability in practice as one essential. In the last chapter I laid down that all elements must be functions of the personality and that they should be conceived of as having dimensions which, in the mind of the analyst, would be sense-impressions, myth and passion.

In this chapter I propose to approach the problem afresh by seeking an answer to the following question: considering any psycho-analytical session as an emotional experience, what elements in it must be selected to make it clear that the experience had been a psycho-analysis and could have been nothing else?

Many features of a psycho-analysis may be regarded as typical but they are not exclusively so. Departures from the common rule of meetings between two people may seem insignificant, but the number of such apparently insignificant departures taken together ultimately amounts to a difference that decides the need for a special term. A catalogue of such difference is likely to establish what constitutes an *imitation* of psycho-analysis rather than what is genuine, unless the difference can be stated in elements.

It would seem to avoid the disadvantages of cataloguing differences in detail by trying to focus on the emotional peculiarities of the experience but difficulty arises because it is common for patients to feel that analysis is coldly unemotional and yet provokes effects appropriate to an intense series of emotions. The surest guide is experience and to this I shall resort in the hope that it can be described in terms that enable others to compare their experience with it.

The dictum that an analysis must be conducted in an atmosphere of deprivation is usually understood to mean that the analyst must resist any impulse in himself to gratify the desires of his analysands or to crave gratification for his own. To narrow the expression of this statement without contracting the area covered by it; at no time must either analyst or analysand lose the sense of isolation within the intimate relationship of analysis.[1]

No matter how good or bad the co-operation may turn out to be the analyst should not lose, or deprive his patient of, the sense of isolation that belongs to the knowledge that the circumstances that have led to analysis and the consequences that may in future arise from it are a responsibility that can be shared with nobody. Discussions of technical or other matters with colleagues or relatives must never obscure this essential isolation.

Opposed to the establishment of a relationship yielding experiences of a sense of responsibility is the drive to be mean and greedy.[2]

The sense of loneliness seems to relate to a feeling, in the object of scrutiny, that it is being abandoned

[1] The point is discussed again under premonition.

[2] Terms such as "greed" are used because I am discussing the elements of psycho-analytic practice. Once such elements are clearly discerned as a part of the emotional experience the analyst can consider in what way they are illuminated by psycho-analytic theories of, for example, anal erotism.

and, in the scrutinizing subject, that it is cutting itself off from the source or base on which it depends for its existence.

To summarize: Detachment can only be achieved at the cost of painful feelings of loneliness and abandonment experienced (1) by the primitive animal mental inheritance from which detachment is effected and (2) by the aspects of the personality that succeed in detaching themselves from the object of scrutiny which is felt to be indistinguishable from the source of its viability. The apparently abandoned object of scrutiny is the primitive mind and the primitive social capacity of the individual as a political or group animal. The "detached" personality is in a sense new to its job and has to turn to tasks which differ from those to which its components are more usually adapted, namely scrutiny of the environment excluding the self; part of the price paid is in feelings of insecurity.

CHAPTER FIVE

The conclusions of the previous chapter suggest that decision requires further discussion; does it involve the translation of thought into action, or some analogous process, for example of thought into a fixed idea, a variable as it were into a constant? Since the analyst is constantly called upon to decide whether to intervene with an interpretation, decision, and its components of loneliness and introspection, should be regarded as an element of psycho-analysis at least from the point of view of the analyst and therefore probably from the point of view of both patient and analyst.

Introspection, that any practising analyst can carry out for himself, into what cliches he most commonly uses often suggests that the problem in analysis is to know which of possible interpretations is at a given time correct; it arises from awareness of the number of ideas expressed in papers on analysis and even more from the variety of human behaviour as it is experienced in ordinary life. In practice the impression is not so formidable: analytic interpretations can be seen to be theories held *by* the analyst about the models and theories the patient has *of* the analyst. It is believed and intended that the analyst's theories, if correct in content and expression, exert a therapeutic effect. Introspection will I believe show most analysts that the theories they employ are relatively few in number and may be seen to fall into the following categories:

1. Definition. Roughly such interpretations take the form that the patient is showing by his associations

that he is, say, depressed. In so far as it is a definitory hypothesis it is by way of saying "This, that you, the patient, are now experiencing is what I, and, in my opinion, most people, would call depression." In so far as it is to define for the patient what the analyst means by definition there can be no argument about it because the only valid criticism would be if the statement could be shown to be absurd because self-contradictory.

2. Statements representing the realization in such a way that the analyst's anxiety that the situation is unknown and correspondingly dangerous to him, is denied by an interpretation intended to prove to himself and the patient that this is not so. Any practising analyst appreciates that this state of affairs belongs to the domain of counter-transference and indicates analysis for the analyst. But as even analysts cannot have all the analysis they may consider desirable the theory used as a barrier against the unknown will remain in the armoury of analyst as well as patient.

3. Statements that are representations of present and past realizations. An example of such a statement would be a brief summary reminding the patient of something that the analyst believes took place on a previous occasion. This corresponds to the function Freud denotes by the term *notation*.[1]

4. Statements representing a scientific deductive system in so far as such a system can be expressed in ordinary conversational English. Such a statement has affinities with 3 above in that it may be regarded as representing a realization from which it has been derived. But essentially its function is similar to that of *attention* as described by Freud.[2] It is the statement one expects to follow an analyst's cliche. "I would like to

[1] *Two Principles* C.P. Vol. IV, p. 15.
[2] *Two Principles* C.P. Vol. XIV, p. 15.

draw your attention to . . ." It is similar to 5 below, but more passive and receptive, corresponding to reverie. It is a theoretical formulation, expressed with as much scientific rigour as the circumstances of analytical practice permit, whose function is to probe the environment. In this respect it has affinities with the pre-conception. It is essential to discrimination. One of its functions is receptiveness to the selected fact. (By selected fact I mean that by which coherence and meaning is given to facts already known but whose relatedness has not hitherto been seen.)

5. Similar to 1, 2, 3 and 4 as far as formulation is concerned—all are formulated by an identical representation, or, in other words, the interpretation can be verbally identical in each case—but it is a theory used to investigate the unknown. The most obvious example of this is the Oedipus myth as Freud has abstracted it to form the psycho-analytic theory. The function of the theoretical formulations in this category are interpretations being used with an intention to illuminate material, that would otherwise remain obscure, in order to help the patient to release still further material. The primary object is to obtain material for satisfaction of the impulses of inquiry in patient and analyst. Note that the probing quality of such interpretations may help to account for differences of reaction in the patient from those he would display to interpretations in category 1 or 4; this component can be distinguished from those derived from the content of the interpretation.

6. In this, the last category that I propose to distinguish, the statement, though still embodied in a representation identical with those employed in all the other statements, is used as an operator. The intention

[1] See Poincaré, H.: *Science and Method*, p. 30, and Hanson, N. R.: *Patterns of Discovery*, p. 121.

is primarily that the communication will enable the patient to effect solutions of his problems of development. (The patient of course can use it to effect solutions of his problems rather than solutions of his problems of development, that is, he can use the interpretation as advice not interpretation, but it is not here my intention to discuss these and other responses of the patient.) Functions of interpretations that fall in this category, and therefore the interpretations in this one of their aspects, are analogous to *actions* in other forms of human endeavour. For the analyst the transition that comes nearest to that of decision and translation of thought into action is the transition from thought to verbal formulations of category 6. From what I have said in Chapter 4 it is clear that activities of this category are those in which the sense of loneliness and isolation are most likely to be in evidence.

These categories are not exhaustive or exclusive. Experience it is to be hoped may lead to substitution by better categories. It is essential to resist an impulse to increase the number of categories unduly partly because that is very easy to do but also because what is needed for my present purpose is the least number of fundamental categories.

I must stress that though in practice interpretations will certainly be embodied in the most diverse formulations, theoretically the same interpretation formulated in the same terms may easily be used in all of these six, and yet more, ways in the same session. The categories I have drawn up to relate, not to the content of theory or the form in which it is represented, but to the work it is intended to do. I shall anticipate by saying that these categories apply to the use to which "thoughts" may be put, once they have been represented by patient as well as analyst. This chapter has been taken up with a particular aspect of what may broadly be called thoughts after the thoughts have been represented by words or combinations of words.

It is therefore a categorization of "*I*" (p. 4, Chap. 1) according to uses to which representations of *I* may be put. This treatment of *I* is a schematic exposition and excludes the component of time that is implicit in a genetic or developmental exposition. In view of the importance that *I* must now assume as a candidate for establishment as one of the elements of psycho-analysis I propose to devote the next few pages to a genetic, as opposed to schematic, exposition of *I* although it involves some repetition of ideas I have already put forward in my paper on Thinking.[1]

[1] "Symposium on Thinking." Int. Congress of Psycho-analysis. Edinburgh 1962.

CHAPTER SIX

The classification I have suggested for analytic interpretations can be applied to all statements whether made by patient or analyst. But I wish to introduce another mode of classification for the same material and for this I propose to draw on experience with patients suffering from disturbances of thought. In contrast with the scheme I have drawn up in the last chapter this will be framed genetically not systematically. Whether there is a realization approximating to it is a question I leave open for the present.

1. β-elements. This term represents the earliest matrix from which thoughts can be supposed to arise. It partakes of the quality of inanimate object and psychic object without any form of distinction between the two. Thoughts are things, things are thoughts; and they have personality.

2. α-elements. This term represents the outcome of work done by α-function on sense impressions. They are not objects in the world of external reality but are products of work done on the sensa believed to relate to such realities. They make possible the formation and use of dream thoughts.

3. I do not consider that there is or can be any evidence for the existence of a realization corresponding to β-elements, α-function, or α-elements, other than observed facts that cannot be explained without the aid of such hypothetical elements. For the remaining formulations the position is different. It can be supposed that there is evidence for the existence of

dream thoughts, preconceptions and the rest. To continue:

3. Dream thoughts. These depend on the prior existence of β- and α-elements: otherwise they require no elaboration beyond that which they have received in classical psycho-analytical theory. They are communicated by the manifest content of the dream but remain latent unless the manifest content is translated into more sophisticated terms.

With dreams one reaches a realm in which there is direct evidence of the phenomena with which one has to deal. At least there is direct evidence when a patient says he had a dream and proceeds to recount it. Unfortunately such assurance evaporates when the subject of investigation is thought itself. The statement that a patient has had a dream is ordinarily sufficient evidence to allow work to proceed, but not if we need to know what has occurred when the patient says he has dreamt. For example, if a patient complains that he had a pain in his leg are we to suppose, in the appropriate setting, that he *dreamt* that he had a pain in his leg or ought we to consider that sometimes the manifest content of a dream is a series of pains rather than a series of visual images that have been verbalized and linked by narrative?

4. The pre-conception.[1] This corresponds to a state of expectation. It is a state of mind adapted to receive a restricted range of phenomena. An early occurrence might be an infant's expectation of the breast. The mating of pre-conception and realization brings into being the conception.

5. The conception. The conception may be regarded as a variable that has been replaced by a constant. If we represent the pre-conception by $\psi(\xi)$

[1] This description of pre-conception is provisional. The concept is elaborated later, notably in Chap. 18 and subsequently.

with (ξ) as the unsaturated element, then from the realization with which the pre-conception mates there is derived that which replaces (ξ) by a constant. The conception can however then be employed as a pre-conception in that it can express an expectation. The mating of $\psi(\xi)$ with the realization satisfies the expectation but enlarges the capacity of $\psi(\xi)$ for further saturation.[1]

6. The concept is derived from the conception by a process designed to render it free of those elements that would unfit it to be a tool in the elucidation or expression of truth.

7. The scientific deductive system. In this context the term "scientific deductive system" means a combination of concepts in hypotheses and systems of hypotheses so that they are logically[2] related to each other. The logical relation of one concept with another and of one hypothesis with another enhances the meaning of each concept and hypothesis thus linked and expresses a meaning that the concepts and hypotheses and links do not individually possess. In this respect the meaning of the whole may be said to be greater than the meaning of the sum of its parts.

8. Calculi.[3] The scientific deductive system may be represented by an algebraic calculus. In the algebraic calculus a number of signs are brought together according to certain rules of combination. The signs have no properties other than those conferred on them by the rules of combination. $(a + b)^2 = a^2 + b^2 + 2\,ab$, is an affirmation of the rules of combination of a and b. a and b have no meaning[4] other than that

[1] Compare with what I say in Chap. 18 on the abstraction.

[2] Compare the state of logical relatedness with what I say in Chap. 18 on coherence.

[3] For a full description of the terms scientific deductive system and algebraic calculus as used in rigorous scientific method see Braithwaite, R. B.: *Scientific Explanation*, C.U.P. 1955.

[4] See discussion of coherence and meaning, Chap. 18.

they are replaceable by numbers and must be understood to be capable of manipulation in the manner defined by the statement $(a + b)^2 = a^2 + b^2 + 2\,ab$. In short to say that *a* and *b* have properties could only mean that they lend themselves to manipulation in accordance with rules and that the rules to which they conform can be deduced from the statement in that, like the conception, it retains a capacity for saturation.

This completes my genetic exposition. I propose now to combine it with the schematic exposition of Chapter 5. This, it will be remembered, adumbrated a provisional scheme by which the various *uses* to which "*I*" might be put could be categorized, thus contrasting with the scheme of the present chapter in which I suggest a scheme of various *stages* by which "*I*" might have developed. It is necessary to note that in the genetic scheme rows B–H inclusive may all be said to contain unsaturated elements that await a realization before they can be "satisfied" and become available for further employment as preconceptions. Row A differs from all others in that it has no unsaturated element and is therefore unsuited for use as preconception. Row B, the α-element, I propose for the present to leave undiscussed for a special reason. For the same reason I shall ignore important aspects of row C, the dream thoughts, and dreams themselves, until later. In the table on the end paper I set out the systematic and genetic expositions along different axes.

This formal table imparts an air of rigidity that may seem alien to a clinical approach. I hope that subsequent discussion involving its use will dispel any fears on that account—provided it is properly used. I shall indicate what that use is by taking up a few of the implications of the grid. The reference numbers are to the co-ordinates in the grid.

A1. This category may be defined as extremely primitive. It shows no clear differentiation of qualities such as we expect to find, say, in a dream as it is recounted by a patient. It shows no differentiation of animate qualities from inanimate, subject from object, moral from scientific. Since it is saturated it is unsuited for use as a preconception. The only sense in which it can be regarded as having a use as a definition is in the sense that to define something may be said to imprison it within certain limits: its meaning is not liberated by verbalization but denied an outlet. It is however suited to projective identification. To use the example I employed in my paper on Thinking—to save unnecessary burdening of the reader's memory by a multiplicity of examples I shall use very few and ask him to bear with the boredom of repetition—the infant experiencing a fear that it is dying, in so far as the terminology of adult sophistication can express the experience at all, imprisons it in a β-element (now placed in the grid category A1). This is projected into the container and its subsequent fate depends on a number of contingencies that I shall not anticipate here because I deal with them later.

A2. The indications I have given in A1 show that strictly speaking A2 must be a nul class, because A1 is incapable of development. Yet in some sense A1 can be used to fulfil some of the functions of A2 in that the imprisonment implicit in A1 denies any liberation of meaning. But comparison with G2 will show that there is a great difference between A2 and G2 (in so far as A2 can be said to exist by virtue of substitution of A1 and A2) and the implication of this difference must be correspondingly greater.

I shall not take up A3, A4 and A5 in detail for what I have said about A2 obtains with appropriate modifications. Essentially they are nul classes. But

A6 is worth brief comment in that the β-element, dealt with by projective identification, does lend itself to use as an operator. Its significance is more sharply defined by comparison with D6, E6, F6, G6 and H6, which I have not so far discussed.

In the situation where the β-element, say the fear that it is dying, is projected by the infant and received by the container in such a way that it is "detoxicated", that is, modified by the container so that the infant may take it back into its own personality in a tolerable form. The operation is analogous to that performed by α-function. The infant depends on the Mother to act as its α-function.

Stating this in other terms, the fear is modified and the β-element thereby made into an α-element. Re-stating this less abstractly still, the β-element has had removed from it the excess of emotion that has impelled the growth of the restrictive and explusive component; therefore a transformation has been effected that enables the infant to take back something, call it an α-element for convenience, that is now suitable for use as a definition or preconception. The change that is brought about by the mother who accepts the infant's fears, is one that is brought about later in personalities whose development is relatively successful, by α-function. By the same token α-function may be described as concerned with the change I have associated with the conception[1] and the concept (E and F in this chapter) as I have described these entities in my genetic exposition.

[1] See commencement of Chap. 6 for discussion in dynamics of growth and Chap. 18 *et seq.*

CHAPTER SEVEN

I SHALL represent the table set out in the last chapter by the sign *I*.[1] I do not propose to discuss what meaning if any is to be attached to classes represented by co-ordinates such as 5.1. We need not suppose that such elements exist. Nevertheless I do not wish to discard them for the present; I propose to reconsider the axes of the framework in the search for elements. When I use the sign *I*, I mean it to represent either the whole table or any one or more of the compartments I have distinguished by co-ordinates. As an example suppose that in the course of an analysis the material suggests the predominance of *I*. This impression should be gained as a result of relaxed or free-floating attention; this state of mind approximates to that represented by D4 (since I am already disposed by my personality and psycho-analytic training to entertain certain expectations). A state of attention, being receptive to the material the patient is producing, approximates to a pre-conception and therefore the change from attention to preconception is represented by a move from D4 on the grid to E4. If I seek confirmation from other material that the patient is presenting, E3 and E5 are swept into activity; if I begin to *verbalize* my impressions F5 is also involved. If it now appears that the time is ripe for an interpretation a further shift takes place, this time towards G6 with a view to a formulation intended to affect the patient.

[1] See Chap. 19 for discussion of feelings.

As aspects of the patient's behaviour relevant to his psycho-analysis fall in the class of phenomena represented by *I* they will be represented by certain of the tabulated categories. Suppose the patient has said, at the commencement of the session, "I know that you do not like me." From knowledge of the patient I may think he is referring to something in the previous session. It would then be a theory enshrining his view of a past event. In that case the realization may be regarded as approximating to G3. But if the G3 session leads me to think that the material is intended by the patient to support the supposition that I do not like him then I would consider that his remark belonged to the category represented by G1 that is to say that it approximated to a definitory hypothesis.

If the context of the statement leads me to suppose feelings of persecution are operating and that his preconceptions would interpret my behaviour as evidence, then his statement would lie in the category E4 and F4. If, however, I regarded it as intended to evoke a confirmation or refutation I would consider it to be classifiable under G6.

I have been supposing in this example that the significance of the patient's behaviour lay in the domain *I*. So far, however, I have in my example been considering the content of his thought in order to determine the category into which, in a given instance, it would fall. If the context of the analysis showed that the content related to an oedipal rivalry the categories of *I* to which it belonged would in most cases be of subsidiary importance, namely for determining the nature of the presenting oedipal material. But if *I* itself is in question the importance of content lies in its relevance to determination of the *I* category. All the categories in the table, with the possible exception of the row B sets, may be considered to play a part, sometimes more important, sometimes less, in any psycho-analytic material. The reader can see for himself that there are some categories in the table into which the analyst's thought processes ought not to fall. Except possibly in the writing of papers or in extra-analytic activities it is difficult to imagine how he could need calculi even if they were available: similarly but for different reasons, familiar to any analyst, he should not be employing any of the column 2 categories. The grid tabulation may help to make explicit features of the analytic situation that should always be watched for as possible disturbances of the analysis.

My immediate concern is with the use of the grid when the presenting problem is *I* itself. The table sets out to cover comprehensively all phenomena that might be described in ordinary conversation as "thoughts" though the right of some categories to be so described might be disputed. As I have said elsewhere[1] patients who suffer from disorders of thought seem to owe their disability partly to failures in development of thoughts themselves, exemplified in the table by the β-elements, and partly to failures in development of an apparatus for dealing with thoughts. It would be easy to say that the obvious thing to do with thoughts is to think them; it is more difficult to decide what such a statement means in fact.[2] In practice the statement becomes more meaningful when it is possible to contrast what a psychotic personality does with thoughts *instead* of thinking them, and how much discipline and difficulty a measure of coherent thinking involves for anyone. I shall ignore any of the uses to which organized thought is put partly because I have already included them as factors in *I*-function and partly because experience in disorders of thought show

[1] Bion, W. R.: *Learning from Experience.*

[2] See Chap. 18 below for mechanisms related to coherence and comprehension.

that their relevance to disorders of thought is chiefly to illuminate them by contrast.

I shall state the theory first in terms of a model, as follows: The infant suffering pangs of hunger and fear that it is dying, wracked by guilt and anxiety, and impelled by greed, messes itself and cries. The mother picks it up, feeds it and comforts it, and eventually the infant sleeps.

Reforming the model to represent the feelings of the infant we have the following version: the infant, filled with painful lumps of faeces, guilt, fears of impending death, chunks of greed, meanness and urine, evacuates these bad objects into the breast that is not there. As it does so the good object turns the no-breast (mouth) into a breast, the faeces and urine into milk, the fears of impending death and anxiety into vitality and confidence, the greed and meanness into feelings of love and generosity and the infant sucks its bad property, now translated into goodness, back again. As an abstraction to match this model I propose an apparatus, for dealing with these primitive categories of *I*, that consists of a container ♀ and the contained ♂. The mechanism is implicit in the theory of projective identification in which Melanie Klein formulated her discoveries of infant mentality.[1] I propose provisionally to represent the apparatus for thinking by the sign ♂♀.[2] The material, so to speak, out of which this apparatus is manufactured is *I*. The material with which this apparatus is designed to deal, is *I*. *I* develops a capacity for any one of its aspects to assume indifferently the function ♂ or ♀ to any other one of its aspects ♀ or ♂. We must now consider *I* in its ♂♀ operation, an operation usually spoken of in ordinary conversation as thinking. From the point of

[1] Klein, Melanie: *Notes on Some Schizoid Mechanisms* 1946.
[2] See Chap. 18 *et seq.* under coherence and comprehension.

view of meaning thinking depends on the successful introjection of the good breast that is originally responsible for the performance of α-function. On this introjection depends the ability of any part of *I* to be ♂ to the other part's ♀. The relevance of this to explanation and correlation I shall deal with elsewhere;[1] briefly, explanation may be seen as related to the attitude of one part of the mind to another, and correlation as a comparison of content expressed by one aspect of *I* to content expressed by another aspect of *I*.

[1] See Chap. 18 *et seq.* under coherence and comprehension.

CHAPTER EIGHT

CERTAIN contradictions and confusions must now be considered even though present knowledge may be inadequate for their resolution. First I propose to review the genetic axis in the light of the aspect of projective identification I have represented by ♀♂. Anticipating what I have to say in Chapter 17, I shall assume that the operation ♀♂ is benign and, as I have suggested, that it is responsible for the developments implied by the genetic ordering of the lettered axis, A to H. (To understand what I mean by the benign operation of ♀♂ see the model on page 35.) Inspection of A to H in the light of ♀♂ shows that the categories have a common relation to each other in that each category depends on changes, in the previous category, that fit it to operate as a preconception as well as a record. Thus E1 depends on mating D1 with a realization that enables the formation of a conception which is in turn capable of leading to F1. Putting this in other terms the element represented by D1 say, is increasing the scope of its function of notation in such a way that its function of attention (the terms "notation" and "attention" being employed in the sense used by Freud[1]) is also increased. Using the table to say this again in a different way, D1 develops through the stages represented by D3 and D4 to become E1.

The mechanics of change from one to another of the phases represented by A to H may thus be represented

[1] Freud, S.: *Two Principles of Mental Functioning.*

by ♀♂.[1] The link ♀♂ between the phases represented by categories A to H, is mechanical. What of the dynamic link? That is represented by L, H and K. The benignity of the operation ♀♂ will depend on the nature of the dynamic link.

The systematic axis of *I*, the uses to which a formulation may be put, consists of a series of categories that could be extended. Since the formulation is the same, only the use that can be made of it is varied; it is obvious that the link between the various categories of use is the formulation. In fact what has to be sought is the counterpart in the systematic axis of the mechanism that links the categories of the genetic exposition genetically. Such a search would involve investigation of mechanisms of evasion and modification of pleasure and pain into which I cannot enter here. It is probable that the mechanism by which transition from one use of the axis 1–6 is transformed into another, is that used in evasion or modification and the dynamic is pleasure and pain.

Melanie Klein's discoveries of the paranoid-schizoid and depressive positions required a theory that in certain situations apparently unrelated elements, associated with feelings of persecution, come together as an integrated whole associated with feelings of depression. I shall employ this theory together with the term "selected fact," borrowed from H. Poincaré.[2] Each "use" classified under the categories 1–6 of the schematic axis depends on the operation of this mechanism on the elements A–G. Thus the use that consists in the employment of an aspect of the categories A to H to inquire or research has come into being by virtue of this mechanism and itself effective by employment of this mechanism. This mechanism I represent

[1] Further consideration will show that this mechanism is related to growth.
[2] Poincaré, H.: *Science and Method*, p. 26. Dover Publications.

by the sign Ps ↔ Dep. The dynamic link, as before, is L, H or K.

The process of change from one category represented in the grid to another may be described as disintegration and reintegration, Ps ↔ D. The benignity or otherwise of change effected by the mechanism ♀♂ depends on the nature of the dynamic link L, H or K.

It will be observed that in the course of the discussion, commenced by making a distinction between thoughts and the apparatus for using them and then according them priority in time so that they could be studied separately from thinking, it has been necessary to reintroduce a primitive mechanics of thinking, or something very like it, to explain the development of the thoughts. In fact it is easier to believe that this spontaneous development in the discussion represents the facts with a greater approximation to the truth than is the case if the accord of priority to thoughts, convenient epistemologically, is to be taken as an accurate representation of the reality of thinking. Nevertheless there are grounds for supposing that a primitive "thinking," active in the development of thought, should be distinguished from the thinking that is required for the use of thoughts.[1] The thinking used in the development of thoughts differs from the thinking required to use the thoughts when developed. The latter is derived from the Ps ↔ D mechanism that is considered in Chapter 9.[2] When thoughts have to be used under the exigencies of reality, be it psychic reality or external reality, the primitive mechanisms have to be endowed with capacities for precision demanded by the need for survival. We have therefore to consider the part played by the life and death instincts as well as reason, which in its embryonic form

[1] See growth and the interaction of ♀♂ and Ps ↔ D in Chap. 18.
[2] And Chap. 18–20.

under the dominance of the pleasure principle is designed to serve as the slave of the passions, has forced it to assume a function resembling that of a master of the passions and the parent of logic. For the search, for satisfaction of incompatible desires, would lead to frustration. Successful surmounting of the problem of frustration involves being reasonable and a phrase such as the "dictates of reason" may enshrine the expression of primitive emotional reaction to a function intended to satisfy not frustrate. The axioms of logic therefore have their roots in the experience of a reason that fails in its primary function to satisfy the passions just as the existence of a powerful reason may reflect a capacity in that function to resist the assaults of its frustrated and outraged masters. These matters will have to be considered in so far as dominance of the reality principle stimulates the development of thought and thinking, reason, and awareness of psychic and environmental reality.

CHAPTER NINE

THE mechanism of projective identification enables the infant to deal with primitive emotion and so contributes to the development of thoughts. The interplay between the depressive and paranoid-schizoid positions is also related to the development of thoughts and thinking. It has been pointed out (by Melanie Klein and Segal) that symbol formation is related to the depressive position. It is compatible with a connection between a capacity for thinking and the interplay between the two positions. It would seem that there is a connection between Ps ↔ D and ♀♂ yet the dissimilarity makes it hard to see what form the connection, if there is one, could take.[1] The bringing together of elements that have apparently no connection in fact or in logic in such a way that their connection is displayed and an unsuspected coherence revealed, as in the example from Poincaré,[2] is characteristic of Ps ↔ D. The operation Ps ↔ D is responsible for revealing the relationship of "thoughts" already created by ♀♂. But in fact it seems as if Ps ↔ D is as much the begetter of thoughts as ♀♂. The development requires examination in some detail.

The earliest observation I have been able to make seemed to suggest that development of thinking through Ps ↔ D depended on the production of signs. That is to say the individual had to bring together elements to form signs and then bring signs together

[1] See below, Chap. 18, in discussion of growth.

[2] Poincaré, H.: *Science and Method*, p. 30. Dover Publications.

before he could think. In his case "writing" preceded not talking only but thinking. His actual speech was incomprehensible if I tried to unravel it by applying my knowledge of ordinary words and grammar. It became more meaningful if I thought of it as a doodling in sound, rather like tuneless and aimless whistling; it could not be described as speech, poetic speech, or music. Just as aimless whistling fails to be music because it does not obey any rule or discipline of musical composition, just as doodling fails to be drawing because it does not conform to the discipline of artistic creation, so his speech for lack of obedience to the usages of coherent speech did not qualify as verbal communication. The words employed fall into an undisciplined pattern of sound.[1] This pattern the patient believed he could see, because the words and phrases that he uttered were believed by him to be embodied in the objects in the room. Having uttered actual objects not phrases the pattern that they formed supposedly revealed their meaning, and this meaning he expected he could now take back. The resemblance this bears to projective identification will be noted.

The procedure I have just described qualifies as an attempt to establish thought, because although the verbalizations seemed to refer to the objects present, and to depend upon their being present, scrutiny showed that the objects were being used as signs to make thinking possible about objects that were *not* present. In this respect the objects in the room were being used as a mathematician might employ mathematical notation to solve a problem without having to rely on the physical presence of the objects on which the problem centred. Ordinarily if a man wished to know how many apples there were if four men had three each he would not need to have the men and the

[1] See below: coherence and meaning.

apples present because he could employ a mathematical notation and the rules for manipulation of its signs. The patient exhibiting the features I have described would, had he been able to employ the objects in the room successfully, have "thought about" objects *not* in the room. It is important to note that in this example the objects in the room are not symbols but signs. In so far as the patient has to wait the appearance of appropriate objects before he can think they are inadequate as signs: in so far as they are not the actual objects about which he is attempting to "think" they represent an attempt to invent and employ signs. To that extent *this* employment of actual objects represents a degree of liberation from the state of mind compelled to employ actual objects.

Amongst these object-signs, as I shall call them for convenience, there is felt to exist one that will harmonize them all: by virtue of its supposed functions it resembles Poincaré's "selected fact." It differs from the selected fact, as I use the term, in that it is not felt by the patient to be other than a thing in itself and this β-element, unlike the selected fact, depends on external fortuitous occurrence.

It is tempting to suppose that the transformation of β-element to α-element depends on ♀♂ and the operation of Ps $\leftrightarrow$ D depends on the prior operation of ♀♂. Unfortunately this relatively simple solution does not adequately explain events in the consulting room; before ♀♂ can operate, ♀ has to be found and the discovery of ♀ depends on the operation of Ps $\leftrightarrow$ D. It is obvious that to consider which of the two ♀♂ or Ps $\leftrightarrow$ D is prior distracts from the main problem. I shall suppose the existence of a mixed state in which the patient is persecuted by feelings of depression and depressed by feelings of persecution. These feelings are indistinguishable from bodily sensations and what might, in the light of later capacity for discrimination, be described as things-in-themselves. In short β-elements are objects compounded of things-in-themselves, feelings of depression-persecution and guilt and therefore aspects of personality linked by a sense of catastrophe:[1] fuller elaboration will have to wait on clinical discovery.

I do not claim the existence of a realization that corresponds to the description that follows; it must be taken as a representation of an hypothesis that is necessary to give coherence to diverse clinical observations.

The β-elements are dispersed; this dispersal should be terminated by Ps ↔ D and a selected fact unless the patient seeks a container, ♀, that compels cohesion of the β-elements to form the contained, ♂.

The dispersed β-elements, in so far as they seek the ♀, may be regarded as an abortive prototype of a container, a container loosely structured like the reticulum[1] of Dr. Jaques. They may equally be regarded as the abortive prototype of the contained, a loosely structured ♂ before compression to enter ♀.

This description can be restated in terms of Ps ↔ D: the cohesion of β-elements to form ♂ is analogous to the integration characteristic of the depressive position; the dispersal of β-elements is analogous to the splitting and fragmentation characteristic of the paranoid-schizoid position.

To restate the above description in more sophisticated terms: the dispersal of β-elements has analogies with the preconception that is to be mated with a realization to produce a conception; the expectation of a breast mating with the realization of the breast.

Although the compression of β-elements to form ♂

[1] There is a curious parallel in a description by R. B. Onians (*Origins of European Thought*, p. 369 C.U.P.) of the Greek ideas of the riddle and the sphinx.

and the dispersal to form a loosely-knit ♀ (a reticulum in search of ♂) is suggestive of Ps ↔ D in fact it cannot be regarded as equivalent because β-elements lack the valency necessary for true integration. The interplay between paranoid-schizoid and depressive positions belong to a stage when the elements can be integrated and the integration represented by verbal statements composed of articulated words. Such statements represent the realization by the nature of their formulation as well as their content. The concentration of β-elements is closer to agglomeration than to integration or coherence: the associated depression and persecution are correspondingly incoherent.

If the dispersed β-elements find no container (the model corresponding to ♀ is presumably the breast) the dispersed β-elements, functioning as we have seen as a loosely-knit reticulum (container in search of a container) become as it were far more actively depressed-persecuted and greedy. The expelling object, the centre of these β-elements, already impoverished by the dispersal, is then threatened with annihilation by its evacuated β-elements since the dispersed elements are searching for saturation. Developments that follow have been described by Melanie Klein and her co-workers and need not detain us now. My main object is to establish the theoretical relationship between the theory of projective identification and the theory of paranoid-schizoid and depressive position.

CHAPTER TEN

In the second paragraph of the last chapter I described behaviour designed to develop thought by the interaction of Ps ↔ D and objects in external reality that were regarded as β-elements. I likened the process to doodling or writing as a method of evacuating objects that could then be scrutinized or dealt with in some way that would cause them to yield a meaning. This process I have described as part of the development of a capacity for thought, the manipulation of β-elements by the mechanism PS ↔ D, may also be regarded as a stage in the development of self-consciousness; for β-elements are felt to contain a part of the personality in their composition. The significance of this will be seen to lie, when we reconsider what the elements of psycho-analysis are, in their supposed possession of characteristics such as greed, love, hate, envy, curiosity. The mechanisms involved in these primitive phenomena can be regarded, at their simplest, as PS ↔ D (or fragmentation ↔ integration) and ♀♂ (or expulsion ↔ ingestion). I shall describe these mechanisms now by reformulations in terms of models.

PS may be regarded as a cloud of particles capable of coming together, D and D as an object capable of becoming fragmented and dispersed, PS. PS, the particles, may be regarded as an uncertainty cloud. These elementary particles may be regarded as closing on to one elementary particle, object, or β-elements, a process that is a particular instance of the general movement represented by → D.

D may be regarded variously as an integrated object,

as an agglomeration produced by the convergence of elementary particles on to one particle or β-element, or as an especial instance of integrated object, namely, either ♀ or ♂. It may even be taken to represent the universe of dispersed fragmentations or elementary particles PS. That is to say, if the *field* of fragmentations is the significant feature then D may represent the whole field of elementary particles.

It will be observed that PS is capable of functioning as if it were a form of ♀. A realization corresponding to this abstraction may be seen in practice when a patient appears to pour out a series of incoherent, inarticulated and disjointed associations designed to evoke from the analyst a statement that is to fulfil the function of one of the following: (1) a selected fact that will give coherence to the whole (an interpretation), (2) meaningful comment from which the meaning will be extracted, (3) meaningful comment on to which the disjointed associations will fasten to destroy the meaning ("So What?" the patient may reply to the response he has evoked from the analyst), (4) a meaningful comment on which the disjointed associations will fasten in order to possess it. (The patient makes apparently no response whatever but subsequently produces the analyst's thought as his own.)

In sum, the two mechanisms can each operate in its characteristic manner or in a manner typical or reminiscent of the manner of operation of the other. The description I have given of PS operating as if it were a form of ♀ may be thought of as representing a situation in which the mechanism PS ↔ D has become arrested at PS, but, in order to maintain its vital function, assumes the mechanical operative quality of ♀♂ and so retains its dynamic quality. Similarly ♀♂ can assume quality of operation characteristic of PS ↔ D.[1]

[1] In particular ingestion of ♂ by ♀, and penetration of ♀ by ♂ replace some of the functions of the selected fact.

In the latter part of Chapter 7 and subsequently I have been concerned with the mechanics of thinking. I proposed that thoughts should be regarded as prior to the apparatus for using thoughts and in the course of the discussion modified this view by suggesting that "thinking" should be used as a term to describe the processes by which thoughts are produced and the processes by which they are subsequently dealt with. If "thinking" is to be used as a term covering the manufacture as well as the employment of thoughts it should be differentiated so that the activities of creation and employment can be considered separately. I then considered PS ↔ D and ♀♂ separately as mechanisms concerned with the elaboration and employment of thoughts. Lastly I have tried to show that PS ↔ D and ♀♂ are not to be regarded as representing a realization of two separate activities but as mechanisms each of which can at need assume the characteristics of the other. In all of this I have concerned myself with content only in so far as it helped to illustrate the mechanisms. Before I turn to content I must point out a difficulty in the use of the term "content." It is clearly appropriate to hypotheses of the kind I have represented by ♀♂. We have already encountered difficulties inherent in the use of terms such as "mechanism" because of the implied model and its unsuitability to convey a meaning when life is an essential element in the meaning that is to be conveyed. The difficulty involved in the use of the term "content" is similar. Although I shall speak of the Oedipal situation as if it were the content of thoughts it will be apparent that thoughts and thinking may be regarded as part of the content of the Oedipal situation. The term "Oedipal situation" may be applied to the (1) realization of relationships between Father, Mother and child, (2) emotional preconception, using the term "preconception" as I have used it here as that which mates with awareness of a realization to give rise to a conception, (3) a psychological reaction stimulated in an individual by (1) above. I trust the context will make it clear in which of these senses I employ the term.

Freud's use of the Oedipus myth illuminated more than the nature of the sexual facets of the human personality. Thanks to his discoveries it is possible by reviewing the myth to see that it contains elements that were not stressed in the early investigations because they were overshadowed by the sexual component in the drama. The developments of psycho-analysis make it possible to give more weight to other features. First, the myth by virtue of its narrative form binds the various components in the story in a manner analogous to the fixation of the elements of a scientific deductive system by their inclusion in the system: it is similar to the fixation of the elements in the corresponding algebraic calculus where that exists. No element, such as the sexual element, can be comprehended save in its relationship with other elements; for example with the determination with which Oedipus pursues his inquiry into the crime despite the warnings of Tiresias. It is consequently not possible to isolate the sexual component, or any other, without distortion. Sex, in the Oedipal situation, has a quality[1] that can only be described by the implications conferred on it by its inclusion in the story. If it is removed from the story it loses its quality unless its meaning is preserved by an express reservation that "sex" is a term used to represent sex as it is experienced in the context of the myth. The same is true of all other elements that lend themselves to abstraction

[1] This point becomes more clear when I discuss the use of ideational content of a statement as a method of expressing feeling. See Chap. 19 *et seq.*

from the myth.[1] In so far as I am concerned to elucidate the elements of psycho-analysis I shall regard the train of causation, as it is expressed in the myth, as an element as we may think it necessary to abstract, but otherwise to be subordinate to the function of linking all elements to confer upon them a particular psychic quality. In this respect the elements suffer modification analogous to that of letters of an alphabet held in combination to spell a particular word. Combination of the elements in the story is analogous to the combination of hypotheses in a scientific deductive system.

The chain of causation is necessary to express the moral system of which it is an integral part. The riddle traditionally attributed to the Sphinx is an expression of man's curiosity turned upon himself. Self-consciousness or curiosity in the personality about the personality is an essential feature of the story: psycho-analytic investigation thus has origins of respectable antiquity. Curiosity has the same status in the myths of the garden of Eden and the tower of Babel—it is a sin. Ignoring the narrative chain of the story except for its contribution to linking the components with each other I isolate the following elements:

1. The pronouncement of the Delphic Oracle.
2. The warning of Tiresias, blinded for his attack on the serpents whose coupling he had observed.
3. The riddle of the Sphinx.
4. The misconduct of Oedipus in arrogantly pursuing his inquiry and thus being guilty of hybris.

Added to these are a series of disasters:

5. The plague inflicted on the population of Thebes.
6. The suicides of the Sphinx and Jocasta.
7. The blinding and exile of Oedipus.

[1] Particularly curiosity—the K link.

8. The murder of the King.

It is noteworthy that:

9. The original question is posed by a monster, that is, by an object composed of a number of features inappropriate to each other.

This concludes my brief review of the Oedipus myth in the light of psycho-analytic theory. I shall consider in what respect it is meaningful to regard the Oedipus myth as an important component of the content of the human mind.

CHAPTER ELEVEN

Discussion of the Oedipus myth as a part of the content of the mind meets with inherent difficulties at the outset. The first is typified by the employment in this context of a locution that implies the model of a container. The second is the peculiar feature of the myth in that the following elements seem to correspond to the numbered axis of the grid.

1. The pronouncement of the oracle defines the theme of the story and can be regarded as a definition, or definitory hypothesis. It resembles a preconception, or an algebraic calculus, in that it is an "unsaturated statement" that is "saturated" by the unfolding of the story; or an "unknown," in the mathematical sense, that is "satisfied" by the story. It is the statement of the theme of the story that is to unfold; the description of the criminal who is wanted.

2. Tiresias may be regarded as representing the hypothesis, known to be false, that is maintained to act as a barrier against the anxiety anticipated as a concomitant of any hypothesis or theory that might take its place.

3. The myth as a whole may be taken as the record of a realization and therefore to fulfil the function Freud attributed to notation.

4. The Sphinx stimulates curiosity and threatens death as the penalty for failure to satisfy it. It can represent the function Freud attributed to attention, but it implies a threat against the curiosity it stimulates.

5. Oedipus represents the triumph of determined curiosity over intimidation and may thus be used as a symbol for scientific integrity—the investigatory tool.

It may seem that I am forcing the myth to fall in with preconceptions of my own, but little ingenuity is required to see these facets of meaning. The classical psycho-analytical employment of the myth sheds light on the nature of L and H links; it is equally illuminating for the K link. Features that can serve as symbols for the mechanics of thinking contribute to my suspicion that it is inadequate to regard the Oedipus situation as a part of the *content* of the mind. I propose to leave discussion of the conception of the mind as having content in abeyance until I have dealt later on with the Oedipus myth in its function as a pre-conception.[1]

I turn now to a clinical experience in which analyst and analysand appear to be speaking the same language, to have many points of agreement and yet to remain without any tie other than that of the mechanical fact of continued attendance at analytic sessions. Progress of the analysis reveals a divergence which I shall sum up as follows:

The analyst is, and thinks he is, in a consulting-room conducting an analysis. The patient regards the same fact, his attendance in analysis, as an experience affording him the raw material to give substance to a day dream. The day dream, thus invested with reality, is that he the patient being extremely intuitive, is able without any analysis, to see just where his difficulties lie and to astonish and delight the analyst by his brilliance and friendliness. The patient reports, and the analyst believes, that he, the patient, has had a

[1] See below, Chap. 15, towards the end, and Chap. 19 on the function of the Oedipus myth as an inborn pre-conception intended to match the realization of the parental relationship.

dream. The patient reports, but does NOT believe, that he has had a dream. The dream, an experience of great emotional intensity, is felt by the patient to be a straightforward recital of the facts of a horrifying experience.[1] He expects that the analyst, by treating it as a dream requiring interpretation, will give substance to his day dream that it was only a dream. In short, the patient is mobilizing his resources,and these include the facts of the analysis, to keep at bay his conviction that the dream not only was but is the reality and the reality, as the analyst understands it, is something to be appreciated only for those elements that are suited to refutation of the "dream."

This account is not of a new theory of dreams, but is a description of a state, seen in an extremely disturbed patient, but probably of fairly common recurrence. The "dream," whether it is correctly described by inclusion in the category of dreams or not, is something that would emerge in the session as a hallucination if the patient's capacity for day dream weakened.

The description fails to bring out one marked peculiarity of such a situation and that is the extent to which analyst and patient agree about the facts. But the agreement on facts is similar to the agreement that two people might have about the disposition of lines, light, and shade in a drawing illustrating reversible perspective–one sees a vase while the other sees two faces: what then are the facts on which both are agreed?

In the example of reversible perspective it may be supposed that it is the actual visual impressions and that the divergence is in the domain of preconceptions. It is possible that this might fairly represent the situation with the patient, but it is a matter that has to be

[1] For the grid categorization of this phenomenon see the later discussion on the application of the grid to feelings: Chap. 19.

determined by clinical observation of each case in which such phenomena appear. I prefer not to lay down a general rule. The principle should be that clinical observation must determine where the intersection of the analyst's and patient's views is.

The significance of the agreement between analyst and patient lies in the fact that the agreement is obvious and obtrusive but the disagreement, which may be just as obtrusive, is by no means obvious. It lies in the use of the agreed facts by the patient to deny what he is convinced are the facts. The conflict between the view of the patient and analyst, and in the patient with himself, is not therefore a conflict, as we see it in the neuroses, between one set of ideas and another, or one set of impulses and another, but between K and minus K (−K) or, to express it pictorially between Tiresias and Oedipus, not Oedipus and Laius.

The common-sense view of mental development is that it consists in an increase of capacity to grasp reality and a decrease in the obstructive force of illusions. Psycho-analysts suppose that the exposure of archaic phantasies to modification by a sophisticated capacity for approximation to a series of theories, that are consistent and compatible with the reception and integration of further experience, is therapeutic in its effects. This supposition cannot stand up to rigorous examination, but has to be received with indulgence to yield meaning of value. Is it possible to find any description in which scientific rigour can make up for absence of indulgence? On the answer to this question depends the possibility of arresting the processes of mental and personality deterioration as those terms are understood medically. The first problem is to see what can be done to increase scientific rigour by establishing the nature of minus K (−K), minus L (−L) and minus H (−H). I shall start by considering the mechanics of thinking. I shall not consider further the ♀♂ mechanism as I do not wish to add anything to what I have said about the denudation of each component by the other. The PS ↔ D mechanism may be briefly dismissed: instead of an interaction involving dispersal of particles with feelings of persecution (Chapter 8) and integration with feelings of depression we have in −PS ↔ D disintegration, total loss and depressive stupor, *or*, intense impaction and degenerate stuporose violence. Although these descriptions of −♀♂ and −PS ↔ D are incomplete they may serve until further experience is forthcoming. I must now consider the nature of the mechanism that I have likened to reversible perspective.

Clinically the picture presented is curiously baffling. There is usually no doubt about the severity of the patient's disability, but it is difficult even for the patient to say why he seeks analysis. It may also be possible at first to underestimate the degree of severity of the disturbance. But before long the lack of contact between analyst and patient and the lack of signs of ordinary conflict begin to build up an unmistakable picture. There is evidence that the patient is a prey to extremely painful emotional experiences: the analyst usually has to rely on the patient's report as the only evidence of these. If they take place in the session the patient invariably has a facile "explanation" of what is taking place. The explanation is often couched in terms that successfully disguise the real nature of the experience. If the patient has been in analysis for some time they are manipulated in such a way that they invite interpretation in terms that the patient has learned to expect from the analyst—the "agreed" intersection is thus maintained. Between analyst and patient there is thus established what I have elsewhere[1] called a contact barrier. Is it possible to glean from the mechanisms involved in this behaviour any material that will throw light on minus phenomena (−L, −H and −K) and incidentally on the problem of establishing the elements of psycho-analysis?

[1] Bion, W. R.: *Learning from Experience.* Heinemann.

CHAPTER TWELVE

THE model of reversible perspective, when applied to the analysis, reveals a complex situation. The patient detects a note of satisfaction in the analyst's voice and responds in a tone conveying dejection. (What was said is irrelevant to our immediate concern.) The patient detects a moral supposition in an interpretation: his response is significant for its silent rejection of the moral supposition. That which makes one person see two faces and the other a vase remains insensible, but in the domain of sense impressions there is agreement. The interpretation is accepted, but the premises have been rejected and other silently substituted.

In any interpretation there is a significant assumption, one being that the analyst is the analyst: this assumption may be denied silently by the patient. Although he appears to accept the interpretation he denies its force by having substituted another assumption. Further associations may show what *his* assumption is.

The debate between analysand and analyst is therefore unspoken;[1] what the analyst says is shown to be agreed by both parties to the analysis, but—it is insignificant. The conflict is therefore kept out of discussion because it is confined to a domain which is not regarded as an issue between the analyst and analysand. The supposition that the analyst is the

[1] The analyst can detect this situation if he uses the grid to categorize interpretations and the responses they evoke and then compares the relationship of the grid categories to each other. See Chap. 20.

analyst and the analysand the analysand is but one of these domains of disagreement that is passed by silently.

Suppose it is brought into the analysis: the patient and the analyst are at once in agreement, for what proposition could be more obvious, or one more loyally accepted by the patient? But the mode of acceptance will indicate that the significance, in the patient's opinion, of the analyst's interpretation lay in the analyst's premises—his false premises. The nature of the falsity is not pressed: it lies in the implication that the premises are of the kind that lead one person to see two faces when he might with equal propriety have seen a vase.

A difference of opinion about the significance of known facts differs from the phenomenon I am describing; on such occasions the difference is overt and concerns the facts. In "reversed perspective" the disagreement between analyst and analysand is apparent only when the analysand appears to have been taken unawares; there is a pause while he carries out a readjustment.

The pause may seem indistinguishable from one that a neurotic patient will make in order to digest the interpretation he has heard. I doubt whether the true nature of the pause can be clinically observed; it may be that an ability to differentiate depends always on long experience of the patient's pauses and the discovery, later rather than sooner, that after many months of apparently successful analysis the patient has gained an extensive knowledge of the analyst's theories but no insight. The pause is not being employed to absorb fully the implications of the interpretation, but rather to establish a point of view, not expressed to the analyst, from which the analyst's interpretation, though verbally unchanged and unchallenged, has a meaning other than the one the analyst intended to convey.

Any growth of insight depends on the extent to which the patient has been forced to digest the interpretation in order to effect the change in his point of view. Such misinterpretation differs from the common run of resistances; the patient will often seize on an ambiguity in the analyst's phrasing or intonation to give his interpretation a slant that the analyst does not intend. The difference is difficult to observe because the patient who reverses perspective also employs the common modes of misinterpretation sufficiently frequently to obscure the more serious condition. He will welcome interpretations of these misunderstandings if they emphasize the deliberate aspects of his contribution; it is a relief to believe that his difficulty is under conscious control.

The following is an example from the analysis of an intelligent man who gave the impression, in session after session, of co-operation that was both friendly and informed—if his responses were not scrutinized too keenly. "My secretary," he said, "complains bitterly of his wife: he says that she does not understand him. He says she complains constantly of him: bitter criticisms and hostility about his lack of understanding, inability to love and so on . . ." In the context of this and similar communications of which his analysis consisted, it was evident that a wide range of transference phenomena might be discussed. In view of the patient's experience of analysis, his undoubted ability and sensitiveness, it would be reasonable to suppose that he would observe a transference relationship between his secretary and himself and that he would be able to understand what his communication of this material would imply about his feelings for me. On several occasions his manner of communicating associations made me suspect that the episodes described might have been invented by him to illustrate a theory of transference that he had picked up in the course of his analysis.

I gave interpretations to this and similar communications, all, as I hoped, being reasonably apposite and given sufficient cogency by the context. His responses were varied; they ranged from an almost stupefied silence to apathetic acquiescence followed by more material—more "free associations." Sometimes he would say that he had been "thinking" during the silence, "about what you said." Sometimes he would disagree with the interpretation, or some aspect of it, and then, as if striving to achieve a solution, would come round to agree that I was probably, no—certainly right. On other occasions, when I felt that my interpretation was one with which he must surely be familiar, he would blandly agree as if it were a cliché that hardly stirred a ripple in his thoughts. It was not till I was able to suggest that he made this class of communication because he felt that the episodes he mentioned were utterly incomprehensible that he made a response showing that this was indeed the case. Such a blank failure to understand would be remarkable at any time, but doubly so in a man who had had so much experience of being analysed. It could not be explained by lack of intelligence, lack of sensitiveness, lack of experience, or ineptitude of my analysis; for the instances he communicated were nearly all of the kind that might be chosen to illustrate psycho-analytic theories.

This last characteristic of these communications is, in this context, particularly baffling: if the patient has no psycho-analytic acumen how is the evidence of careful selection in accordance with psycho-analytic principles to be accounted for? If the evidence of selection is admitted how is one to explain the failure of comprehension?

I exclude the hypothesis of deliberate conscious or semi-conscious denial of the analyst's work. My reason, of which I shall say more later, is the evidence of pain. There was ample evidence, after interpretations had established the reality of the patient's inability to comprehend, of the severity of his pain.

In each instance, the perspective that enabled me, but not the patient, to grasp the meaning of the associations, was that afforded by the Oedipus theory. In every instance, that which appeared to cause the patient to reverse the perspective, was the Oedipus myth. I say myth, not theory, because the distinction is important: the Oedipus theory and its various formulations, belongs to the area of the grid, covered by F4, G4, F5, G5. The myth pertains to the area C.

The patient's capacity for learning but not using analytic theories is a failure to match pre-conceptions with the realizations that approximate to them. The unsaturated element remains unsaturated.

Melanie Klein described a situation in which the personality attacked its object with such violence that not only was the object deemed to become minutely fragmented, but the personality likewise. In the situation I describe there appears to be no dynamic splitting. It is as if the splitting was arrested in a static pose action being no more necessary than it is when hallucination is substituted for reality. The patient does not have to disagree with the analyst or experience Oedipal conflicts within himself: he reverses the perspective. It is important to consider more closely what this means.[1]

Reversing perspective is not the same as evacuation of β-elements. That is an active process and the patient's behaviour in the analysis affords evidence

[1] The "static" splitting and failure to match pre-conception with realization will be better understood when it is related to discussion in first half of Chap. 19.

that is, to put it no higher, quite compatible with a theory that he is taking action to "disburden his psyche of accretions of stimuli" as Freud described it.[1] As near as I could tell the patient regarded the analyst's interpretations as evidence that he, the patient, had evacuated β-elements, a state of mind more akin to hallucination than delusion. The search for elements involves further investigation of pain, reversible perspective and the Oedipus myth.

[1] Freud, S: *Two Principles of Mental Functioning.*

5—20 pp.

CHAPTER THIRTEEN

Reversible perspective is evidence of pain; the patient reverses perspective to make a dynamic situation static. The work of the analyst is to restore dynamic to a static situation and so make development possible. As I said in my last chapter, the patient manoeuvres so that the analyst's interpretations are agreed; they thus become the outward sign of a static situation. Since the analyst's interpretations are unlikely to permit this and the patient is unlikely always to command sufficient nimbleness of mind to match the interpretation with a shift that reverses the perspective in which the interpretation is viewed, the patient employs an armoury that is reinforced by delusion and hallucination. If he cannot reverse the perspective at once he can adjust his perception of the facts by mis-hearing and misunderstanding so that they give substance to the static view: a delusion is in being.

If this is not sufficient to keep the situation static the patient resorts to hallucination. For simplification, I can restate this as hallucination in order to preserve, temporarily, an ability to reverse perspective; and reversed perspective in order to preserve a static hallucination.

The prolonged resort to reversible perspective is thus accompanied by delusions and hallucinations that are difficult to detect because they are both static and evanescent. Moreover, since their aim is to preserve the analyst's statements (interpretations) as an overt expression of agreement and a defence against change, the true significance of the patient's behaviour as a sign of delusion or hallucination is not apparent unless the analyst is alert to this possibility. The analyst's expressed thoughts belong to the area F5, G5 and G6: the same formulation is accepted by the patient but as an expression of thought belonging to F1, G1, G2.

An analysis that is taking this course appears to be curiously unsatisfactory because the lack of real progress obtrudes only slowly and then appears to be stable, unspectacular and dull. In reality the situation is unstable and dangerous. The clue to it lies in the fact I mentioned at the start of this chapter—pain. The patient's manoeuvres appear to lack an aim because, though the ready acceptance of interpretations arouses suspicion, it is not apparent that they are directed against change, *any* change, and pain. It is the dynamic quality of the interpretation that evokes evasive reactions. That is to say, the objection to the interpretation is that, *whatever* the content may be, it has the qualities appropriate to columns 5 and 6 of the grid.

Reference to the grid would lead one to suspect that if the patient manoeuvres to shift all *I* to the area of columns 1 and 2, with the possible addition of 3, he must tend to do the same with his own *I* phenomena. That is in fact the case and it helps to explain some of the features of his dreams, pre-conceptions and theories.

The lesson to be drawn from this discussion is the need to deduce the presence of intense pain and the threat that it represents to mental integration. I shall therefore consider pain as one of the elements of psycho-analysis.

Pain cannot be absent from the personality. An analysis must be painful, not because there is necessarily any value in pain, but because an analysis in which pain is not observed and discussed cannot be regarded as dealing with one of the central reasons for the patient's presence. The importance of pain can be dismissed as a secondary quality, something that is to disappear when conflicts are resolved; indeed most patients would take this view. Furthermore it can be supported by the fact that successful analysis does lead to diminution of suffering; nevertheless it obscures the need, more obvious in some cases than in others, for the analytic experience to increase the patient's *capacity* for suffering even though patient and analyst may hope to decrease pain itself. The analogy with physical medicine is exact; to destroy a capacity for physical pain would be a disaster in any situation other than one in which an even greater disaster—namely death itself—is certain.

In reversible perspective acceptance by the analyst of the possibility of an impairment of a capacity for pain can help avoidance of errors that might lead to disaster. If the problem is not dealt with the patient's capacity to maintain the static situation may give way to an experience of pain so intense that a psychotic breakdown is the result.

The case for acceptance of pain as an element of psycho-analysis is reinforced by the position it occupies in Freud's theories of the pleasure-pain principle. It is evident that the dominance of the reality principle, and indeed its establishment, is imperilled if the patient swings over to the evasion of pain rather than to its modification; yet modification is jeopardized if the patient's capacity for pain is impaired. As I have discussed the relationship of the establishment of the reality principle to psychosis in *Learning from Experience* I shall say no more about it here.

Pain cannot be regarded as a reliable index of pathological processes partly because of its relationship with development (recognized in the commonly used phrase "growing pains") and partly because intensity of suffering is not always proportionate to the severity of the disturbance. Its degree and significance depend on its relationship with other elements.

Implicit in the discussion of reversible perspective as a means of preserving a defence against pain is the concept of growth. Growth is a phenomenon that appears to present peculiar difficulties to perception either by the growing object or the object that stimulates it, for its relationship with precedent phenomena is obscure and separated in time.[1] Difficulty in observing it contributes to the anxiety to establish "results," e.g. of analysis. It will be necessary to trace its relationship with PS ↔ D, ♀♂. Its dependence on a capacity to entertain the social and narcissistic components of the Oedipal situation involves further discussion of the Oedipus myth, the Babel myth (Genesis XI, 1–9) and the early version of the Eden myth (Genesis II, 8–3 passim). The primitive models for mental growth are the Tree of Knowledge, the Tower and city of Babel and the Sphinx. The myths (Row C of the grid) provide a succinct statement of psycho-analytic theories which are relevant in aiding the analyst both to perceive growth and to achieve interpretations that illuminate aspects of the patient's problems that belong to growth.

[1] One of the advantages of reference to the grid is that grid categorization of the patient's response to interpretation should reveal growth.

CHAPTER FOURTEEN

In Chapter 3 I described personal myth as an important tool in psycho-analytic work. In chapters 11 and 12 I attributed augmented though similar significance to the Oedipus myth because it has public and racial, as opposed to private, status. The advantages of transition from private to racial myth are analogous to the transition from private to public communication.[1]

The Oedipus myth will be differently read by different people, but the measure of agreement makes it a channel of public communication as Freud's use of it has shown. I shall use the myths of the garden of Eden and the Tower of Babel to reinforce the expression, already implied by the Sphinx in the Oedipus myth, of attitudes of a god inimical to the gaining of knowledge by human beings whose search is felt to imperil his supremacy.

In the garden of Eden, possessed by the Father, eating of the Tree of Knowledge of good and evil is forbidden. The serpent, or Satan disguised, incites the woman to defy the ordinance of the Almighty. The relevation of disobedience is associated with guilt and nakedness. The outcome is banishment as it is in the Oedipus myth. In the Babel myth the use of the tower is to effect an entry into realms regarded by Jahweh as his own—heaven. The outcome is exile, as in the garden of Eden and Oedipus myths, but an important precursor is the destruction of a common language and

[1] Compare *Scientific Explanation*—R. B. Braithwaite, p. 6, and his reference to Heinrich Hertz on p. 91.

the spreading of confusion so that co-operation became impossible.

The components of these myths that I wish to use are those which pictorialize, in the sense of internal pictures or symbols that we make for ourselves, features that might turn out to be the psycho-analytical elements that I seek.

1. There is a god or fate, omniscient and omnipotent though modelled anthromorphically. This god belongs to a moral system and appears to be hostile to mankind in his search for knowledge, even moral knowledge.

2. In all, penetration into, or ingestion into, or expulsion from, a blissful place or state is prominent. Sexual knowledge and pleasure is a prominent feature of the knowledge sought and forbidden.

3. In Eden and Oedipus myth there is a stimulation of forbidden desires—the serpent incites desire for the fruit: Oedipus instigates the search for the criminal: in the Babel myth there is a significant variation—the people come together and are dispersed, the language is one and is replaced by a number of languages. The Sphinx incites to curiosity by its riddle. Using these myths as a source for pictorial representation of the elements in the horizontal axis of Table 1 the definitory function of a formulation may be represented by the oracle and the statement of aims as expressed in the phrase "let us build us a city and a tower." The repressing force exerted by the formula used in col. 2, by Tiresias or the god of fate of whom he is the representative. The action column 6 is represented by the outcome, exile or dispersion.

It is not my object to establish an exact correspondence; I suggested in Chapter 11, that these myths serve as a primitive counterpart of the sophisticated formulations whose employment in scientific work I have placed on the horizontal axis of the grid. Since they are primitive and pictorial they are vital but lack precision—hence the need for sophisticated formulations as in science. Therefore to make the correspondence between the horizontal axis and the elements of the myth appear to be exact would be a falsification that obscured the nature of the myth. Equally failure to see that there is this correspondence obscures the value of the myth as a fact-finding tool. I wish to restore it to its place in our methods so that it can play the vitalizing part there that it has played in history (and in Freud's discovery of psycho-analysis) and it is for that reason that I have introduced it in Chapter 3. It is also an object for investigation in an analysis as part of the primitive apparatus of the individual's armoury of learning.

If the Oedipal myth in addition to the place it already occupies in analytic theory is recognized as an essential part of the apparatus of learning in primitive stages of development, a number of elements discernible in the debris of a disintegrated ego assume a new significance.

In certain highly disturbed patients the patient according to Melanie Klein attacks his object with such violence that not only is the object felt to disintegrate but the personality delivering the attack also. This disintegration is characteristic of the patient who cannot tolerate reality and has therefore destroyed the apparatus that enables him to be aware of it. The private myth, corresponding to the Oedipus myth, enables the patient to understand his relationship with the parents. This private myth, in its investigatory role, if impaired or maldeveloped or subjected to too great a stress, disintegrates; its components are dispersed and the patient is left without an apparatus which would enable him to comprehend the parental relationship and so to adjust to it. The Oedipus debris will in these circumstances contain elements that are components of the Oedipal myth that should have operated as a pre-conception. How are the scattered components of a disintegrated ego to be recognized? In this instance the analyst who seeks to illuminate the fragments of the patient's apparatus of learning may be led to recognize them by noting isolated[1] fragments of the Oedipal myth (and the myths I have associated with it).

The private myth has this important role in the individual's attempt to learn from experience analogous to that played by public myths as systems of notation and record in the development of groups. Conviction comes only from clinical experience in which material resembling Oedipal components appears, the components being dispersed and evocative in tendency. The myth must be expected to appear in a private version. The operation follows the pattern of the fragments in PS $\leftrightarrow$ D as I have described it at the commencement of Chapter 10. The verbalizations of the patient, and behaviour associated with them, seem to vary from something incoherent and meaningless to formulations that appear vaguely to be inviting comment, sometimes even suggesting the comment that they invite.

Search for the elements of psycho-analysis is restricted to that aspect of them that it is the business of the psycho-analyst to discern. They cannot be represented, either by abstract signs, such as I have suggested, or by mythological narratives evoking visual imagery, in such a way that anyone other than a trained and practising psycho-analyst could recognize the realization approximating to the representation. In this chapter I hope that by showing a correspondence

[1] Such fragments appear, particularly in psychotic material, widely dispersed in analytic time. One of the problems of interpretation is to show that these temporarily dispersed fragments are related.

between myth and the items of the horizontal axis of the grid analysts, who are used to observing patients in the light of the preconceptions with which psycho-analytic Oedipal theory endows them, will find it easier to make the transition from theoretical back-ground to consulting-room phenomena. The gap remains and can only be bridged by training and experience and, to some extent, by interpolation, be-tween the concept of "Oedipal components" and the events of the consulting room, of description of those events in terms that are vivid and exact. Such descrip-tion would tax the powers of a writer and would then represent a degree of particularization unsuitable except for short and relatively infrequent passages of a total analysis.

Although I aim to isolate elements of the reality of psycho-analytic practice and not of theory I have to represent them by signs and myths that belong to the domain of representation of abstractions and the high-level hypotheses of a scientific deductive system. This must not obscure the fact that every sign is intended to represent phenomena that a psycho-analyst might experience in analytic sessions. The signs chosen to represent the elements are to aid in working on and thinking about the experience of the analysis.

CHAPTER FIFTEEN

In this chapter I reconsider transference. The elements of the transference are to be found in that aspect of the patient's behaviour that betrays his awareness of the presence of an object that is not himself. No aspect of his behaviour can be disregarded; its relevance to the central fact must be assessed. His greeting, or neglect of it, references to couch, or furniture, or weather, all must be seen in that aspect of them that relates to the presence of an object not himself; the evidence must be regarded afresh each session and nothing taken for granted for the order in which aspects of the patient's mind present themselves for observation are not decided by the length of time for which the analysis has endured. For example, the patient may regard the analyst as a person to be treated as if he were a thing; or as a thing towards which his attitude is animistic. If ψ (ξ) represents the analyst's state of mind *vis-à-vis* the analysand it is the unsaturated element (ξ) that is the important one in every session.

The peculiarity of a psycho-analytic session, that aspect of it which establishes that it is a psycho-analysis and could be nothing else, lies in the use by the analyst of all material to illuminate a K relationship. The transference interpretation is peculiar in that it refers to all material without discrimination, but is highly selective in appraising its significance. The patient communicates information that has significance by virtue of criteria of his own: the analyst is restricted to interpretations that are an expression of a K relationship with the patient. They must not be expressions of L or H.

Ordinarily an analyst would not expect to go through a series of conscious speculations on the nature of simple statements. But if he wishes to do some "homework," that is to say extra-analytic meditation, on a session, either to train and amplify his capacity for intuitive deduction, or because he has some doubts about the accuracy of the work he has been doing, he may refer the doubtful material to the grid. Suppose the patient had said "I know that you hate me": his speculations might take some form such as the following: On the face of it the sentence is grammatically and semantically straightforward, but there is need to examine it critically to determine which of the grid categories apply.

The context of the statement, taken together with the (ξ) element of ψ (ξ), representing the patient's state of mind, could make it probable that the significance lay in its being for the patient a disguised expulsion of flatus: in that case the analyst would assess the chance that it was a β-element pertaining to row A.

The session might indicate that the phrase related to a dream that the patient had had or was perhaps part of a phantasy. In that case it would belong to rows B or C.

In yet another context it might be more appropriate to suspect that it was a pre-conception and should lie in the class represented in row D.

Suppose however that the statement had followed a critical interpretation that could itself have been interpreted by the patient as an expression of hostility: it might then be appropriate to regard it as, from the patient's point of view, a mating of a pre-conception with a realization.

It cannot be supposed to lie in row H; it might, however, be correctly attributed to row G, but for this to be appropriate the analyst would need to be able to draw on a number of attendant circumstances that led him to suppose that the statement represented a hardening of belief into a more nearly permanent fixed idea.

Let us now consider the horizontal axis. The columns in this axis represent the functions that a statement is being made to perform. The statement may be an oracular pronouncement, an announcement of the theme of the session, a definition in the light of which the remainder of the session is to be understood. In short it may fall in the category represented by column 1.

If it seems rather to fall in the column 2 category it will mean that the statement is known to be false but provides the patient with a theory that will act as a defensive barrier against feelings and ideas that might take its place.

If it seems to fall in columns 3, 4 or 5, it is a probing statement and is quite compatible with co-operation in the analytic investigation.

If it belongs to column 6 it is a warning of acting out, in which I include the use of the analysis itself as a form of acting out. To illustrate the value of the grid as an instrument to aid the analyst in thinking about an analytic problem, that is to say, as an instrument of notation that provides a record of fact and a sign that can be manipulated in a manner analogous to a number in mathematics, I shall contrast A6 and F6 by considering the significance of each.

A.6 indicates that the statement "I know that you hate me" is to be regarded as a β-element employed as an act. If the analyst concludes that the statement falls in this category it can only mean that the session must be regarded as an acting out rather than as an ordinary analytic session. The muscular movements necessary to express the words, since they are intended to disburden the psyche of accumulations of stimuli, must be regarded as of primary significance. The words indicate that the patient has a feeling (which is apparently regarded by him in much the same light as a normal personality would regard a concrete object) that was part of his personality (also regarded as a concrete object) that he can by his muscular movements detach and expel. The end-product of this manoeuvre is that he has achieved a state of mind in which he is no longer burdened by the feeling that the analyst hates him. He is now, supposedly, free to feel that the analyst is his friend.

To turn now to F6. If this is the category to which the statement belongs it now means that the patient is convinced that the analyst is his enemy. It means furthermore, by virtue if its membership of column 6, that the patient is already acting, or is about to act, on this supposition. The analytic significance of the patient's statement is very different in the two instances; it is a matter of importance to distinguish between the two categories and to decide to which of the two the statement belongs.

Consider the statement "I know that you hate me" as it appears in another category; suppose that it belongs to the categories lying in row C, because the patient said that he had dreamt it. Or it might be part of a phantasy or day dream: in such a case the characteristics of visual imagery, or perhaps myth, would be prominent. The intensification of this element is bound to affect the elements expressed by the headings of the horizontal axis of the grid. Such interplay is implied by the grid. Consequently the columns 1–6 tend to become pictorialized and personified in the way I have suggested by equating each of them with a character from the Oedipus myth. For example, if the statement appears to be appropriate to C6 when a break in analysis is impending the break would be related to exile and the analyst should anticipate the appearance of other features of the Oedipal situation. I stress the importance of recognizing the category to which material belongs as a step towards anticipating and therefore recognizing related phenomena in the material. If the material fits category C5 one would expect an intensification of the determination with which curiosity was pursued: if in category C2 an intensification of resistance to the emergence of new material. I shall leave the significance of the emotion expressed—"I know that you *hate* me"—to the next chapter.[1]

The description I have given is not of a process of thought appropriate to actual contact with the patient. The analytic session is too precious an opportunity for observation to be jeopardized by pre-occupations of the kind my description implies; the theme of this book is a scheme to facilitate thinking about analytic work *outside* the sessions themselves. The object of such extra-sessional work is to substitute creative thinking for laborious and frequently meaningless note-taking; it provides practice, analogous to the musician's scales and exercises, to sharpen and develop intuition. It becomes increasingly possible to arrive at conclusions instantaneously which at first are the fruits of laborious intellectualization.

[1] And in the latter part of Chap. 19 where I discuss feelings.

CHAPTER SIXTEEN

Problems of instinct and emotion belong to the main body of psycho-analytic theory and must be considered for inclusion amongst the elements of psycho-analysis as they appear in psycho-analytic practice.

The emotion to which attention is drawn should be obvious to the analyst, but unobserved by the patient; an emotion that is obvious to the patient is usually *painfully* obvious and avoidance of unnecessary pain must be one aim in the exercise of analytic intuition. Since the analyst's capacity for intuition should enable him to demonstrate an emotion before it has become *painfully* obvious it would help if our search for the elements of emotions was directed to making intuitive deductions easier. The sexual instinct is an integral part of psycho-analytic theory, but the element of sex in the sense of something for which I need to look is not sex but that from which the presence of sex may be deduced. But for my purpose the term "element" cannot be properly used to denote something that would appear to be a property of some more fundamental thing whose presence it betrays. Therefore the element I choose is not a sign of sexuality but a precursor of sexuality. Amongst the elements that we seek there must be the precursor of emotion, not an emotion itself unless it is the precursor of some emotion other than itself. Thus, if the hate that a patient is experiencing is a precursor of love its virtue as an element resides in its quality as a precursor of love and not in its being hate. And so for all other emotions.

I am adumbrating, in the domain of emotion, something that is reminiscent of the relationship of preconception to conception.

If interpretations forestall the development of emotions by illuminating their precursors it follows that sexual and other feelings should not be regarded as elements. The counterpart of the preconception is the *premonition.* Directly observed emotional states are significant only as premonitions.

I have defined the pre-conception as an element private to the individual, possibly not conscious; the same is true of the pre-monition.

Lest it seems confusing to use the term preconception as something to be distinguished from a scientific deductive system and then speak of an analytic theory as an analyst's preconception I shall use the term pre-conception to distinguish it from a preconception. Pre-conception, as I have placed it in row D of the grid, is a term representing a stage in the development of thinking; preconception, in the sense of the analyst's theoretical preconceptions refers to the use of a theory and so belongs to columns 3 and 4 of the grid. As an example, the analyst may have a hunch that something of which he is not properly aware is going on in the session. This state might be represented by D3. As he feels greater conviction about the development, his state of mind changes until, say, he conceives that oedipal material is emerging; his state of mind is now represented by G4 or G5. In other words he thinks that Freud's oedipal theories will give point and direction to his probing curiosity. To return to the significance of emotional drives that are active in the course of the analytic experience and the manner of their elucidation; when a patient comes for a first consultation his premotions give information about him that cannot be obtained from other factors. From them it is possible to obtain some idea of the use that he is likely to make of an analysis.

The term "premonition," as I propose to use it represents emotional states rather than ideational content, thus leaving the term "pre-conception" to represent the latter. I do not dissociate "pre-monition" from its association with a sense of warning and anxiety The feeling of anxiety is of value in guiding the analyst to recognize the premotion in the material. The premonition can therefore be represented by (Anxiety (ξ)) where (ξ) is an unsaturated element.

Analysis must be conducted so that the conditions for observing pre-monitions exist, a conclusion compatible with Freud's definition of the analytic situation as one in which an atmosphere of deprivation[1] is dominant. If premonitions cannot be experienced correct interpretation becomes difficult for the analyst to give and difficult for the analysand to grasp; unnecessary pain, of which I have already spoken, becomes more likely.

The distinction between pre-conception and premonition facilitates the creation of a system for thinking about analytic practice; it is no more of a falsification than the separation, implied in the use of the term "sex" or "fear," of one emotion from another. Freud showed that in certain hysterical palsies and anaesthesiae the distributions corresponded to *ideas* of anatomical structure, rather than to the nerve distribution known to an anatomist. An anatomist can use the concept of a hand provided he does not allow his view of anatomical structures to be obscured by using it in investigations to which it is inappropriate. A patient says he feels fear but does not know of what he is afraid. We also suppose that man can feel fear but not sex. What is the model implicit in a statement

[1] See Chap. 4, p. 16 above.

that a man feels afraid, but does not know why? One possibility is that it is derived from the model of himself that the parents have presented him with. It would seem to suggest that "feeling afraid" and "knowing why," or "sex" and "fear" are different things. Correct treatment depends on the view we take of a patient who has lost all feeling "in his mind." By analogy it is important to know what the premises are on which we base a view that "sex" and "fear," or "feeling afraid" and "knowing why" differ. The supposition that pre-conception and premonition are to be distinguished in this universe of discourse and its implication that there is a similarity between the objects thus distinguished leads me to the point I wish to establish, namely that the categorization of ideational content that is facilitated by the grid is equally significant if it is applied to the emotional experience. For example we may regard the statement "I feel that you hate me" is significant for the idea it expresses. This idea can then be placed in one of the grid categories. Or we may regard the statement as significant for its expression of an emotion. The emotion (or pre-monition) can then be placed in one of the grid categories. The matter will be discussed more fully after we have considered the problem of "feelings" which is raised in Chapter 19.

CHAPTER SEVENTEEN

THE grid itself can obviously be categorized in accordance with its own categories. Thus the horizontal axis may be described as a series of definitions of various uses. In so far as it is employed to define the horizontal axis belongs to the categories of column 1. But as it is, its use could qualify it for inclusion in row F. But suppose it were desired to test out the value of some of the suggestions made in this book; in that case the horizontal axis might be regarded as a pre-conception to which one wished to find a matching realization. As the object of investigation it would fall in the category of a bound constant conjunction, a definitory hypothesis that stated that certain elements are constantly conjoined. As the object of further investigation it would then be treated in a manner intended to invest it with meaning or expose whatever meaning it might have.

As I have set it out in the grid the horizontal axis is represented by abstract signs, in a manner, as I said above, that can qualify it to belong to one of the categories of row F. But, following a suggestion implicit in the grid, I shall take the "uses" of the horizontal axis and substitute personifications, thus incidentally restating it in terms that would qualify it for categorization in row C.[1]

It might be convenient to substitute any sign or

[1] The preceding suggestions are examples of the use of the grid to choose a mode of employment for a concept. It contrasts with the use of the grid to find the category for a psycho-analytic object that has been employed already.

symbol that I chose, for example, the sign ψ for column 2. By proceeding in this way I can fill the line with signs and symbols that have meaning for me and thereby establish the basis for private communication, that is, for communication private to myself. Instead of the sign ψ I can use the sign "Smith" a man known to me who has a significance private to myself. Since however this discussion is intended to be public, I shall employ signs representing symbols that are public already and are consequently more likely to effect a public communication. This I can do by using elements in a myth familiar to the audience I hope to reach. It must be appropriate to the matter to be symbolized on the one hand and to culture of the group to which I want to address myself on the other. As this is addressed primarily to psycho-analysts I shall use the Oedipus myth, in the way indicated at the end of Chapter 15, the Jehovistic account (the second in Genesis) of the Creation and the account in Genesis XI of the building of the Tower and City of Babel. Both accounts are Jehovistic and anthropomorphic, the latter characteristic contributing to their value for this discussion.

Since I am using a myth as the course of my symbols the substitution is itself an artificial[1] use of row C elements rather than those of, say, G or H. I can choose characters to correspond to the separate columns 1, 2, 3, etc., or I can place the entire myth in each of the compartments C1, C2, C3, etc., or any single character, symbolizing 1–6 can be placed in *all* the columns 1–6 in row C. Thus Tiresias, symbolizing col. 2, may now appear in cols. 1, 3, 4, 5, 6. If it is used in column 2 it would represent a symbol representing an idea entertained for the purpose of denying the emergence of a more accurate but more frightening idea. In this

[1] See previous note.

instance it would represent the adherence to a myth of a repressing force, for use as a repressing force.

Since Tiresias is a symbol for the use represented by col. 2 it may seem that it would be shorter and simpler to say that any use represented in cols. 1–6 may be placed in any other column—and its own. But in fact this is not so: a use cannot act as a use; C2 cannot be meaningful placed in C2.[1] But a symbol representing a mode of thought, as "Tiresias" represents dream or myth or model, can, by being placed in column 2, represent a dream or mythological thought used to inhibit another mode of thought even though this other mode of thought is being used to inhibit. Thus a deductive system used to inhibit the emergence or other thoughts may itself be inhibited by a dream of mythological thought used for that purpose. Tiresias, or a corresponding private symbol, may be used to inhibit the use of a scientific theory to inhibit further thoughts. Stating this abstractly, in terms of the grid, C2 can be used to inhibit G2.

The reformulation of the uses to which thought can be put by the substitution of symbols drawn from myth has expressed the uses in terms that bring them into the category designated by row C. The new formulation should therefore make it possible to subject the reformulated statement to the treatment or process, whatever it is, that governs the transition of the elements in the vertical axis from one to the next below—a process I have described as one of growth, positive or negative. The uses, formulated in terms of myth, row C, may now be successively diminished in quality till they become analytical objects represented by the

[1] "C2" represents an abstract statement. But the category C2 is intended to contain dream thought not sophisticated statements. C2 in C2 would therefore mean either that C2 or what it represents was wrongly categorized, or, what comes to the same thing, C2 only appears to be a scientific concept but is a myth.

β-elements of row A, or be stimulated to grow so that they can be represented by signs appropriate to the elements of rows D, E, F, G and H.

The categorization from which we started, namely that expressed by the signs used to represent the uses to which a statement may be put (the numbers representing the columns) can itself be seen to belong to row G. The numbers are used purely as a means of notation. They can be seen therefore to be classifiable under column 3.

To substitute personifications as symbols for the "uses" could be described as effecting a transition in the vertical axis from below upwards—from row G to row C—a matter of manipulation of signs and topographical distribution in the grid.

If the total myth is placed in each of the compartments of row C, the compartments represent a primitive scientific instrument[1] for scanning the analytic material. It may itself be represented by F or G5 an instrument for a macroscopical view of the analytic material. If only one component of the myth is used, instead of the whole myth, the grid reading represents an instrument affording a more restricted view of the material—a view analogous to microscopic scrutiny. The movement → A diminishes sophistication of the component represented, but the movement → H increases it, the latter being approximate to a prelude to interpretation.

The grid as the representation of an instrument used by the analyst in scrutinizing the patient is equally a representation of the material produced by the patient as an instrument for scrutinizing the analyst. But if the analyst scrutinizes the material (the realization) to see in what grid category the representation

[1] Cf. Chap. 19. The use of the Oedipal myth as a pre-conception intended to mate with the parental realization to produce understanding of the parental relationship.

lies to which the realization approximates the grid is an instrument and not merely its representation.[1] The realization to which his attention (Col. 4) is directed is the reality of pre-conception and premonition.

So far I have been concerned with manipulations of signs in the grid. Is it possible to say that these manipulations correspond with the dynamics of the realizations represented by the readings in the grid? There is no difficulty in supposing that growth takes place but are the processes of growth, as deduced from observations in the consulting room, approximate to the rules of manipulation of the signs in the grid? The movements can be said to represent the results of growth or diminution—I am not discussing acquisition and denudation related to greed and envy—but for the present the actual manipulation of symbols must not be taken to represent growth or diminution itself. This matter can be reconsidered after discussing the vertical axis.

In the Jehovistic version of the creation a feature of the myth of the expulsion is the apparent conflict between the thirst for knowledge and the will of the deity. In the story of Babel, the god is also opposed to the will of the people; it appears to impinge on his right to occupy heaven undisturbed. The god in both is anthropomorphic in the manner typical of J sources. The Eden shows him opposed to eating that confers a knowledge of good and evil: the Babel myth shows him opposed to language because the common language confers on the people capacity to co-operate in building both a city and a tower, the latter facilitating entry into the god's heaven. The punishment in Eden is expulsion from the garden: in the Babel story the integrity of the language is destroyed, each fragment becoming a new language, confusion supervenes

[1] See Hanson, N. R.: *Patterns of Discovery*, p. 100, §5 (*d*).

and the differing language groups are scattered. The exile theme common to both stories is discernible in the exile of Oedipus. In all three sex is implied. Knowledge is related to eating in Eden and to morality in so far as it makes discrimination between good and evil possible. In Babel knowledge seems to be referable to scientific rather than to moral standards though the god's possession of heaven is a "moral" issue.

The elements in each of these three myths bear a resemblance to the elements of the other two; from them symbolic representations of oral sexuality and scattering, repressive super-ego, linkage through language, learning and self-knowledge, genital sexuality (e.g. tower and city) may be readily obtained. The difference in the apparent content is due to the form of narrative by which the elements are linked in each story. The relative importance of the elements will depend on the nature of the exploration for which they are used and the fact selected to bring coherence to the elements when they are re-integrated in the process of analysis. Re-integration is not something occurring once for all; throughout an analysis elements of the analytic material are seen by the analyst to be inappropriately assembled and a fresh integration and cohesion is made possible by the elucidation his interpretations afford. The selected fact that gives coherence, can be an idea or it may be an emotion. What emotions give rise to the patient's integrations and disintegrations must be deduced from inspection of the premonitions.

CHAPTER EIGHTEEN

THE vertical axis (A–H), related to a genetic rather than a systematic exposition, involves a premiss of growth dependent on (*a*) psycho-mechanics, (*b*) an alternation of particularization and generalization, (concretisation and abstraction), (*c*) successive saturations, and (*d*) emotional drives.

(*a*) The relationship between the mechanisms of projective identification and the alternation of paranoid-schizoid and depressive positions in K presents difficulties that seem due to an incompatibility. A solution may be approached through investigation clinically of the destructive splitting attacks that transform ♂ into fragments which nevertheless retain in their fragmented form an association with each other sufficient to permit *penetration* of a problem. Similar fragmentation of ♀ leaves an association of fragments that still perform the function of ingesting or introjecting. The objection to attributing priority to splitting lies in the fact that it does not allow primary quality to the alternation between paranoid-schizoid and depressive positions: yet both projective identification (♀♂) and paranoid-schizoid ↔ depressive positions must be regarded as potentially primary.

(*b*) The "alternation between particularization and abstraction" as a method of describing a theory is open to the objection that "abstraction" is a term that implies removal of a quality from something. The theory is more likely to match realizations if the *formulation* of an abstraction or generalization is seen to be the significant feature of the transaction and not extraction of qualities from a known representation or its corresponding realization. Generalization (or abstraction) must be recognized as a process by which the saturation of an unsaturated element is bound in order to consolidate the gain. The abstraction, or formulation, of a generalization consists in the naming[1] of a new entity. What has been regarded as a dynamic state in which elements of a realization are abstracted selectively to form an abstraction, generalization or, more abstractly still, an algebraic calculus, should be regarded as the mating of a pre-conception with a realization to form a conception and *thus* a reformulation: the reformulation is a *naming* of the total constellation of pre-conception and conception to prevent the loss of the experience by dispersion or disintegration of its components. The process known as abstraction is related to notation (as described by Freud) and an enlargement of memory. It is relevant here to consider in more detail the idea of growth positive and negative (Chap. 17, p. 5).

I introduce the idea of negative growth as a method of approaching an aspect of learning from experience; I do *not* mean denudation which I associate with hostile and destructive impulses such as envy. Denudation implies impoverishment of the personality. What I mean is exemplified by the restatement of the horizontal axis of the grid in mythological symbolism rather than in the terms appropriate to a deductive system (row C rather than row F or G). A capacity for negative growth is needed partly to revivify a formulation that has lost meaning, partly to establish a link in making

[1] Compare Condillac's theory that ideas become fixed by being associated with a sign or word: (Etienne Bonnet de Condillac. Essay on the origin of Human Knowledge): Hume on constant conjunction, and Freud on thinking "became endowed with further qualities which were perceptible to consciousness only through its connection with the memory traces of words." (Freud, S.: *Two Principles of Mental Functioning*. 1911.)

private knowledge public, but perhaps most important of all to achieve naivety of outlook when a problem is so overlaid by experience that its outlines have become blurred and its possible solutions obscure. One of the advantages of the grid is that its use in thinking about material that emerges in psycho-analytic practice stimulates reconsideration of familiar phenomena such as dreams or Oedipal material and their corresponding psycho-analytical theoretical formulations. The ability of an analyst to retain the substance of his training and experience and yet achieve a naïf view of his work allows him to discover for himself and in his own way the knowledge gained from his predecessors.

(*c*) The theory implicit in the representation of a pre-conception by a constant ψ together with an unsaturated element (ξ) has its convenience provided it is remembered that the sign ψ (ξ) is a representation of a complex realization. We do not know the nature of the process of saturation or how the extent of the suffusion of the psyche by the stimulus of a new experience is to be determined. I find it a useful representation till it can be discarded for a better.

(*d*) To what I have already said about emotional drives I add a reminder that the analyst's concern is with the premonitory aspects of these drives and that the political nature of the human being should be borne in mind in assessing the force and direction of these premonitions. The determining factors in even intimate manifestations of sex or aggression may lie outside the personality and within the group.[1]

A vertical axis related to a genetic exposition depends on a concept of growth; in the formulation of this axis I have been governed by the idea of pre-conception as central. Hitherto it has served to suppose that abstractions and generalizations are extracted or

[1] Freud, S.: *Instincts and their Vicissitudes.*

abstracted from an already existing concept.[1] With patients in whom disorders of thought are prominent this view of abstractions and generalization fails to explain the nature of his thoughts. The fault lies in the model implicit in the term "abstraction." The term I require must express in the domain of psycho-analysis what is expressed in mathematics when it is said that a formula, already discovered, has been, and may at some future date be, approximated to by a realization. This meaning is inherent in the term "pre-conception" as I wish to use it. It is the meaning I wish to express in terms such as "generalization" and "abstraction." The scientific deductive system (G) and the algebraic calculus (H) may also share the quality of a pre-conception; different terms in the vertical axis express differences in degrees of sophistication rather than differences in function. Stating the same thing in another way terms in the vertical axis vary but all have the same use whereas in the horizontal axis all terms are the same but the "uses" vary. The significance of the formulation I am putting forward lies in its use to establish the theory that any term such as "dog," "unconscious," "dream," "table," comes into existence when a set of phenomena are recognized as having a coherence of which the meaning is *unknown*.

An object is not perceived and given the name of "dog" because from the object perceived a quality of "dogginess" is abstracted. The term "dog" ("unconscious," "dream," "table," etc.) is used when and because a set of phenomena is recognized as being related yet *un*known. It is used to prevent the scattering of the phenomena. Having found the name, and thereby bound the phenomena, the remainder of history, if so wished, can be devoted to determining what it means —what a dog *is*; the name is an invention to make it

[1] See Chap. 1.

possible to think and talk about something before it is known what that something is. To postulate an α-element or α-function as I have done, is, according to this theory, merely a conscious sophisticated extension of normal spontaneous procedure that has always existed and is inherent in the development of language. The term "dog" or the term α-function comes into existence, the one spontaneously and unconsciously, the other by premeditation and contrivance, because phenomena are meaningless and need to be bound together so that they can be thought about. Once a name has been given and the scattering thereby prevented, meaning can begin to accumulate. How readily and spontaneously this can occur is illustrated by the observation[1] that attempts to contrive a term devoid of a distorting penumbra of associations is often defeated by the speed with which such a meaningless term collects a meaning. To sum up: the pre-conception awaits its realization to produce a conception: the term "dog" waits for a real dog to provide it with meaning. The algebraic calculus awaits a realization to approximate to it. In this way mathematicians who say mathematics have no meaning are justified. The mathematical formulae are analogous to pre-conceptions, as I use the term, and await a realization that approximates to them before they can be said to have meaning. The grid, as I have adumbrated it here, itself shares the qualities I attribute to the pre-conception.

It will be observed that this theory of the name as that which prevents scattering of phenomena so that they can function as a pre-conception, is at variance with the theory, that I have already used, that Euclidean geometry, for example, is abstracted from the realization of space and finds therefore its realiza-

[1] Popper, K. R.: *The Logic of Scientific Discovery*, Chap. II, 9 and 10.

tion in space. The theory of pre-conception I am putting forward requires a readjustment in our views about concretization, particularization and β-elements The term "pre-conception" is ambiguous because it denotes a tool, the function for which it exists and the use to which it may be put; the two last may of course be the same.

The genetic view of an abstraction differs from the conception of an abstraction as a representation of something that has been removed from something. A readjustment of our idea of the concrete is required. The term "dog" and all others terms apparently possessing a definite meaning must be seen to have no meaning until they acquire one through accretions of experience. A theory that such terms must be regarded as binding phenomena helps to dissipate contradictions inhering in the theory that an abstraction is something that has been removed from something else. Hume's postulation of constant conjunction is quite compatible with a theory that the term "dog" comes into existence in order to be a sign that certain discrete and previously incoherent phenomena are constantly conjoined. There is no difficulty if this fact is kept clearly separated from any idea that the term implies that the constantly conjoined elements have any meaning other than that the elements are constantly conjoined. The genesis of the term "dog" is thus adequately accounted for as a product of the PS $\leftrightarrow$ Dep mechanism. But only after the term "dog" has served to signalize and perpetuate the conjunction does the question of meaning arise. It might be stated thus: "These phenomena are constantly conjoined. I record this fact and bind the phenomena to each other so that they shall remain constantly conjoined by the sign 'dog.' Now that I have bound the phenomena I can try to find out what their constant conjunction means. Thus we shall arrive at the *meaning* of the term 'dog'."

The mechanism ♀♂ comes into play and the product of its operation is meaning.

The nature of the relationship between PS ↔ Dep and ♀♂ thus receives some clarification. On the PS ↔ Dep operation depends the delineation of the whole object: on the successful operation of ♀♂ depends the meaning of the whole object.

CHAPTER NINETEEN

The choice of axes may appear arbitrary without further reasons; it stems from the analytic situation itself.

The patient's activity most in evidence in an analysis is thinking. The analyst can see the use he makes of the analytic situation. He may appeal for help, exploit the possibilities of cruelty to the analyst, seek an outlet for love and generosity and so on. This he does by thinking silently, talking to the analyst, thinking aloud and occasionally by action. He makes statements classifiable under the headings of the vertical axis A–H. The information available for communication is selected, withheld, or expressed according to the use to which he wishes to put it. Since this is so the significance of the content of the communication can be assessed from the use, only *one* of the features of the patient's contribution, but it is a feature that is most continuously significant and is therefore worthy of attention, isolation and elevation as the 1–6 axis of the grid.

Since self-knowledge is an aim of psycho-analytic procedure the equipment for attaining knowledge, the function and apparatus of pre-conception, must be correspondingly important. Growth and a capacity for growth are equally fundamental. The vertical axis (A–H) represents both the stages of growth and the function of pre-conception.

Representation of a process, such as growth continuing, by the headings A–H gives a misleading impression of discrete, clear-cut entities; transition from one to another must be assumed, for lack of evidence, probably to be gradual.

Choice of the Oedipus myth as a reservoir on which to draw for symbols to replace the horizontal axis (1–6) enables me to illustrate a feature of the myth as a pre-conception. The Oedipus myth may be regarded as an instrument that served Freud in his discovery of psycho-analysis and psycho-analysis as an instrument that enabled Freud to discover the Oedipus complex. I turn now to the part played by the myth, or its counterparts in α- and β-elements, in the growth of the psyche.

The myth may be regarded as a primitive form of pre-conception and a stage in publication, that is, in communication of the individual's private knowledge to his group. Any scientific theory must ultimately be represented by a medium that facilitates publication. The processes by which private knowledge is communicated within the individual is obscure and elucidation waits on advances that psycho-analysts have yet to make. The dream has fresh significance if it is regarded as a private myth. Oedipal material in a dream requires the accepted classical theory that it displays the Oedipus situation as yielded up by the unconscious under the impact of the analytical investigation, but in certain cases this material must also be regarded as evidence of a primitive mechanism of pre-conception, a private version of what later becomes publicly communicable through its correspondence with the Oedipal myth. When the issue is capacity for thought the same is true of the Eden and Babel myths. Analysts need accordingly to consider that the Oedipal material may possibly be evidence for primitive apparatus of pre-conception and therefore possessing a significance additional to its significance in classical theory. I am postulating a precursor of the Oedipal situation not in the sense that such a term might have in Melanie Klein's discussion of *Early Phases of the Oedipus Complex*, but as something that belongs to the ego as part of its apparatus for contact with reality. In short I postulate an α-element version of a private Oedipus myth which is the means, the pre-conception, by virtue of which the infant is able to establish contact with the parents as they exist in the world of reality. The mating of this α-element Oedipal pre-conception with the realization of the actual parents gives rise to the conception of parents.

If, through envy, greed, sadism or other cause, the infant cannot tolerate the parental relationship and attacks it destructively, according to Melanie Klein the attacking personality is itself fragmented through the violence of the splitting attacks. Restating this theory in terms of the Oedipal pre-conception: the emotional load carried by the private α-element Oedipal pre-conception is such that the Oedipal pre-conception is itself destroyed. As a result the infant loses the apparatus essential for gaining a conception of the parental relationship and consequently for resolution of Oedipal problems: it does not fail to solve those problems—it never reaches them.

The significance of this for practice is that scraps of what appear to be Oedipal material must be treated with reserve. If the evidence is related to a disaster to the ego, the destruction of the pre-conception and consequently of the ability to pre-conceive, interpretations based on the supposition that fragmental Oedipal material is evidence of a destroyed object will be only partly successful. The investigation must be directed to distinguishing amongst the elements of Oedipal material those that are fragments of Oedipal pre-conception from those that are fragments of the fragmented Oedipal situation. Since the experience of learning from which the patient is thus debarred is that of the parental relationship, the importance for the patient's development and for a successful outcome of analysis, depending on resolution of the Oedipus complex, are gravely prejudiced.

I must leave the development of this theme for future treatment when I hope to show in greater detail the manner of using the grid for penetrating to a clearer perception and understanding of clinical material. The nature of the elements is determined by their position relative to the two axes (1–6) and (A–H).

What might be called the psycho-mechanics of thinking, represented by the interplay between ♀ and ♂ and between paranoid-schizoid and depressive positions (PS ↔ Dep and the selected fact) provides the link between row and row.

At the beginning of this chapter brief reasons were given for the choice of "uses" for the axis 1–6. In the genetic axis the psycho-mechanisms represented by PS ↔ D and ♀♂ were accorded an importance that requires further discussion. At the end of Chapter 18 I said that when hitherto unrelated elements were discovered, by the operation of PS ↔ D, to be coherent their relatedness was fixed by nomination. The name had a function, analogous to that of the mathematical formulation, in fixing the constant conjunction that it represented. In this respect the fixing of the cohering elements is identical with the mating of preconception with realization to produce a conception. We must now consider the process by which the name accumulates meaning through the operation of ♀♂.

I have discussed the part played in the understanding of thinking by the model of the alimentary canal. Detailed discussion of the mechanisms and dynamics involved in the growth of meaning is helped by introducing a shift of emphasis in our scrutiny of the phenomena represented by the vertical (A–H) axis of the grid. I shall represent this shift of emphasis by using the term "feeling" instead of the term "thinking". This substitution is based on the common usage, in analytic practice, of phrases such as "I feel I had a dream last night," or, "I feel you hate me," or "I feel that I am going to have a breakdown." Such locutions imply an emotional experience and are therefore more appropriate to my purpose than the more austere implication of "I think . . ." Communications introduced by terms such as "I feel . . . ." are often methods of expressing emotions or premonitions. It is in their function as expressions of emotion that I wish to consider these phenomena. I propose to leave the grid unaltered; the categories represented by the co-ordinates of the grid apply equally to classes of "I think . . ." and "thoughts" as to classes of "I feel . . ." and "feelings." In order to indicate emphasis of the emotional content I shall speak of "feeling" rather than "thinking," but the grid remains unchanged for the categorization of "thoughts" or "feelings."

Regarding the statements whose development is represented by the vertical (A–H) axis as expressions of feeling, the mechanism ♀♂ by which change from one row of the grid to another is effected, may be represented by models other than that provided by the digestive tract. Of these the most suggestive are (1) the respiratory system, with which is linked the olfactory system; (2) the auditory system with which is linked transformations such as music ↔ noise, and (3) the visual system. Each of the three provide models for the mechanism ♀♂ representing projective identification as employed for purposes of K. The sense of touch is usually employed as an antidote to the confusion that can be incidental to the employment of ♀♂. Its use to establish the reassurance obtained from feeling there is a barrier between two objects, a limiting boundary that is absent in the container ↔ contained relationship characteristic of models (1), (2) and (3), produces the paradoxical effect that the topographically closer relationship implied by tactile contact is *less* intimate, i.e. confused, than the more distant relationship implied by the models (1), (2) and (3).[1] It is worth noting that clinical manifestations of asthma become psycho-analytically more meaningful if their relationship to the respiratory model for thinking-feeling is recognized.

Views of thinking and feeling, stated in terms appropriate to row G, can be restated, by use of these models of digestive tract, respiratory system, auditory system and visual system in terms appropriate to row C and vice versa.

The mechanism PS ↔ D allied with the mechanism ♀♂ is thus responsible for growth of pre-conception either in the direction of naivety or of sophistication.

Since we can use the grid categories to represent feelings much will depend on the analytic context in which statements are made. The analyst must decide whether the idea that is expressed is intended to be an instrument whereby feelings are communicated or whether the feelings are secondary to the idea. Many subtle expressions of feeling can be missed if the ideas by which they are expressed are regarded, wrongly, to be the main burden of the communication. The facility with which subtle shades of feeling can be expressed makes the communication of what appear to be ideas an ideal vehicle for the communication of premonitions; the "ideas" should be scrutinized accordingly.

[1] A psychotic patient can have genital intercourse without confusion but will become seriously confused by (1), (2) and (3).

If the grid categories are as appropriate to "feelings" as "ideas" there should be an emotional counterpart of β-elements. In the limited field to which I have restricted them hitherto I have suggested that the term "β-elements" should be used to cover the area of phenomena such as the "thoughts" that some psychotic patients regard as indistinguishable from "things." In the domain of feelings and those aspects of thought in which feelings are dominant the term "β-elements" should be extended to cover the analogous phenomena. I am not sure what these analogous phenomena, if any, are. But the same patients who regard "thoughts" as "things" show every sign of regarding what I am used psycho-analytically to believe are phantasies as "facts." I therefore suggest provisionally that the β-element categories of the grid should not be dismissed off-hand as non-existent, but should be thought of, in the domain of expressions of feeling, as related to phantasies that are felt to be indistinguishable from facts. These phantasies that are indistinguishable from facts must be considered as the emotional counterpart of β-element "thoughts" that are indistinguishable from "things." In other words the β-element categories of the grid should be regarded as representations to which some realization, discovered psycho-analytically, may be found to approximate.

CHAPTER TWENTY

BEFORE I summarize the main themes of this book I must make it clear that the grid, though genetically related to a number of analytic theories, differs from the object which the term "theory" is usually taken to represent. Its nature is better indicated by describing it as a convention for construing psycho-analytical phenomena.[1] But if an analyst uses this convention he entertains a pre-conception of which the grid, as printed or written, is a representation. Thus the analyst's state of mind, which is approximating to the representation, the grid as printed or written, can be classified under one of the grid categories according to the use to which it is being put and the position it occupies in the genetic development of the analyst's scientific equipment. Since the grid and the associated theories may be used for whatever motives the analyst chooses it can be used to keep knowledge at bay (column 2) as well as to further it (cols. 1, 3, 4, 5, etc.). Equally, since use of the grid means that the analyst entertains a preconception, it is possible that his use of it can be categorized more precisely if we employ the vertical axis for this purpose and see the appropriate grid category in row D. If, therefore, the reader does not wish his state of mind to be disturbed he will either not use the grid or employ it as a means of expressing feelings and thoughts that can be represented by D2. If on the contrary he is disposed to employ the grid for further investigation he will be employing it as a means

[1] But note the discussion at beginning of Chap. 17.

of expressing feelings and thoughts that can be represented by D (or, according to the degree of sophistication he brings to bear, F (or G)4.

I am aware that the grid not only can be but requires to be improved. I have felt that col. 2 might be replaced by a negative sense to the horizontal axis. It is plausible and would conform to a pleasing similarity with the system of cartesian co-ordinates as used in the development of algebraic geometry. Furthermore, it would simplify some difficulties if, instead of the present arrangement, the horizontal axis read

$$-(n), -(n-1) \ldots -5, -4, -3, -2, -1, 1, 2, 3, 4, 5, \ldots (n-1)\ (n)$$

with the 2 column standing for what is, in the present grid, column 3. Then one might say that all the "uses" $1 \leftrightarrow n$ can be used negatively, as a barrier against the unknown or known but disliked. But I am primarily concerned to sketch out the uses to which the phenomena represented by the vertical axis are put and not to embark on the complexities of the "uses" to which "uses" may be put, regarding the latter as an extension best left to clinical investigation.

I turn now to uses of the grid. The summary is not intended to be exhaustive.

A. Meditative Review. 1. Suppose that at the end of the day's work the analyst wishes to review some aspect of his work about which he is doubtful. Assume further that the preoccupation centres on some phrase of the patient's. Recalling the session, the context of the statement, the patient's intonation, the analyst can place the statement in a category which, in the light of after knowledge, he thinks is correct. Such meditation is related to notation and memory. It is akin to recording what took place and is an example of using the grid and the theories it represents for the purposes of notation. Even if he does not commit his work to paper the analyst is doing something that will stamp the episode in his memory.

2. The analyst may place it speculatively in any grid category he chooses. He can then give direction to his speculations by considering what the implications would be if in fact the statement belonged to the category in which he had provisionally and speculatively placed it. This means that he has "bound" a number of elements and can proceed to discover the meaning of their supposed conjunction. The grid assists in giving direction to his speculations.

3. In the course of 1 and 2 he will be considering the possibility of other categories in which the statement might with propriety have been placed. Such activity is a stimulant to the analyst's capacity for attention.

4. The analyst can scrutinize his interpretations by subjecting them to the same procedure as that to which he has submitted the patient's associations in 1, 3, and 4.

5. The analyst can place the association and its actual, or proposed interpretation, in the appropriate categories and thus examine the *couple*, association and interpretation. Thus he can compare and examine the relationship not of the association to the interpretation but of the category of the association to the category of the interpretation. Thus a basis can be provided for investigation of the developmental value of interpretation and association according to the nature of the relationship of their categories.

6. The analyst can take conflicting statements in the patient's associations, place them according to their respective grid categories, and then scrutinize the nature of the conflict by a comparison of the *categories* of the conflicting statements. It should then be

possible to see what is contributed to conflict by the nature of the categories of the conflicting statements.

B. The psycho-analytic game. In A I have proposed uses for the grid closely associated with actual analytic experiences. The grid may, however, be profitably used in a kind of analytic make-believe in which the experiential element is far less dominant. Such an imaginative exercise is closer to the activity of the musician who practises scales and exercises, not directly related to any piece of music but to the *elements* of which any piece of music is composed. This brings me back to the elements of psycho-analysis and their elucidation. I define the elements of psycho-analysis as being those phenomena whose various aspects can be seen to fall within the grid categories even though some categories must for the time being remain empty. Such phenomena are—

(*a*) Ideas as described in Chapters 1–18.

(*b*) Feelings as described in Chapter 19, including pain.

(*c*) Association and Interpretation.

(*d*) The couple (association and interpretation).

(*e*) Conflicting pairs (I use the term "pair" so as to leave the term "couple" free for the phenomena of (*d*) above).

(*f*) The two axes of the grid (as special cases).

In Chapter 3 I suggested that the psycho-analytic object had three "dimensions," sensa, mythology and analytic theory. Translating this into terms of grid categories any analytic object before qualifying as such must display features categorized in rows B, C and G. An analytic object is not the same as an element but may be regarded as having a relationship with an element analogous to that of a molecule to an atom. The analytic object is not necessarily an interpretation though an interpretation is an analytic object. An interpretation must be based on evidence of analytic objects and is itself an analytic object composed of analytic objects. The analytic object emerges as a result of the operation in the observer of PS ↔ D and ♀♂. To the analytic observer the material must appear as a number of discrete particles unrelated and incoherent (PS ↔ D). The patient may be describing a dream, followed by a memory of an incident that occurred on the previous day, followed by an account of some difficulty in his parents' family. The recital may take three or four minutes or longer. The coherence that these facts have in the patient's mind is not relevant to the analyst's problem. His problem—I describe it in stages—is to ignore that coherence so that he is confronted by the incoherence and experiences incomprehension of what is presented to him. His own analysis should have made it possible for him to tolerate this emotional experience though it involves feelings of doubt and perhaps even persecution. This state must endure, possibly for a short period but probably longer, until a new coherence emerges; at this point he has reached → D, the stage analogous to nomination or "binding" as I have described it. From this point his own processes can be represented by ♀♂—the development of meaning. It has been necessary to give this somewhat schematic description of the analyst's mental working to introduce a discussion of certain apparent anomalies to which I now turn. I shall take the statement I made (Chap. 17) that the whole Oedipus myth could be placed in a single category, or, alternatively, that parts of the myth could occupy a single compartment in the grid. It may seem that this is not compatible with the distinction I have made between a psycho-analytic object and an element of psycho-analysis. The anomaly disappears however if it is appreciated that, in the context in which the

Pre-conception,
as stage of development of thought, 23
Prediction,
need to detect precursors of emotion to avoid unnecessary pain, 74
use of grid to aid prediction, 73
Premonition,
analogy with pre-conception, 75
as term to represent precursors of emotional states, 76
Proclus,
quoted by *Sir T. L. Heath*, 2 (footnote 1)
Progress,
lack of analytic, related to dangerous unstable condition, 61
Projective identification,
and ♀♂, 31
and patterns of sound, 38
and PS ↔ D, 41
related to β-elements, 26
PS ↔ D,
and ♀♂ in operations of the analyst, 102
as begetter of thoughts, 37
contemporaneous with ♀♂, 39
establishment of meaning for -PS ↔ D, 52
mechanism of, 34–35
PS capable of functioning as, ♀, 42
related to fragmented private Oedipus myth, 67
relationship with ♀♂ and constant conjunction, 94
Publication,
analogous with transition from private to public myth, 64

R,
reason, notation for, 4
as slave of the passions, 36
Results,
of psycho-analysis and difficulty of detecting growth, 63
Reversible perspective,
as model for type of disagreement between analyst and analysand, 50
Reversible perspective (*contd.*)
in practice, 54-59
in terms of grid categories, 61

Scientific deduction system,
as logically linked statements, 24
Schematic exposition,
combined with genetic exposition, 25
Scientific procedure,
in psycho-analysis requires myth, 12
Segal, H,
on symbol formation and depressive position, 37
Selected fact,
described by *Poincaré, H*, 39
can be idea or an emotion, 83
Sensa,
related to α-elements, 22
relationship with mythology and theory, 101
Sense impressions,
agreement on, contrasted with disagreement on their interpretation, 54
Sex,
or aggression, determining factors of, may be in group, 86
Signs,
possibly precede thinking, 38
used to make thinking possible about objects not present, 38
Sphinx,
and growth, 63
riddle of, 46
stimulating curiosity, 48
Stages,
of development of thought axis in genetic exposition, 25
Super-ego,
repressive, and deity in myths, 82–3
Symbols,
use of private, and public, 79
Symposium on thinking,
International Congress of Psycho-Analysis, 1962, 21

Systematic Axis,
transformation from 1–6 related to pleasure and pain, 34

Theories, psycho-analytic,
criticized as unscientific, 1
Things,
and thoughts equated, 22
Thought,
and thinking, relative priority, 35
extra-sessional, about analytic work, and use of grid to aid it, 73
failures in development of, 30
ordered, in contrast with psychotic, 31
translation into action; or of variables into constants, 17
Tiresias,
as symbol for barrier against anxiety, 48
warning against inquiry, 45
Transference,
some aspects of, and K, 69
use of grid to represent, 70

Uses,
as axis in a systematic exposition, 25

Verbal communication,
contrasts with incoherence, 38
Vertical axis,
of grid, and growth, 86
of grid, justification for, 92

Wisdom, J. O.,
An Examination of the Psycho-Analytical Theories of Melancholia, 1 (footnote 1)